The Last Writings of Thomas S. Kuhn

THE LAST WRITINGS OF THOMAS S. KUHN

Incommensurability in Science

Thomas S. Kuhn

Edited by Bojana Mladenović

The University of Chicago Press

Chicago and London

The University of Chicago Press, Chicago 60637
The University of Chicago Press, Ltd., London
© 2022 by The University of Chicago
Published 2022
Paperback edition 2024
Printed in the United States of America

33 32 31 30 29 28 27 26 25 24   1 2 3 4 5

ISBN-13: 978-0-226-82274-7 (cloth)
ISBN-13: 978-0-226-83331-6 (paper)
ISBN-13: 978-0-226-51630-1 (e-book)
DOI: https://doi.org/10.7208/chicago/9780226516301.001.0001

Library of Congress Cataloging-in-Publication Data

Names: Kuhn, Thomas S., author. | Mladenović, Bojana, editor,
    writer of introduction.
Title: The last writings of Thomas S. Kuhn : incommensurability in science /
    edited with an introduction by Bojana Mladenović.
Description: Chicago : University of Chicago Press, 2022. |
    Includes bibliographical references and index.
Identifiers: LCCN 2022006350 | ISBN 9780226822747 (cloth) |
    ISBN 9780226516301 (ebook)
Subjects: LCSH: Science—Philosophy. | Science—History.
Classification: LCC Q175 .K938 2022 | DDC 501—dc23/eng20220421
LC record available at https://lccn.loc.gov/2022006350

♾ This paper meets the requirements of ANSI/NISO Z39.48-1992
(Permanence of Paper).

For Sarah Kuhn

CONTENTS

EDITOR'S ACKNOWLEDGMENTS

My work on preparing this volume for publication was greatly helped by many people and institutions. First of all, I am grateful to Karen Merikangas Darling of the University of Chicago Press for inviting me to edit Thomas Kuhn's unfinished book, and for giving me the time and freedom necessary to complete this task in a manner that seemed best to me. I am also grateful to Karen for introducing me to Kuhn's literary executors, his widow Jehane Kuhn and his daughter, Sarah Kuhn. Their enthusiasm for and faith in the project were of vital importance throughout. Jehane's knowledge of the history of the manuscripts and of Kuhn's own ideas concerning posthumous publication of his works was invaluable. She threw light on a number of points that could not be resolved on the basis of the extant texts alone. I am deeply saddened that she did not live to finally hold this book in her hands. Sarah Kuhn, in continuing as her father's literary executor, has displayed outstanding generosity and integrity. I am grateful for her reliable help, and for her genuine warmth. This volume is dedicated to her.

For two semesters, the Oakley Center for the Humanities and Social Sciences provided a peaceful space and research funds to work on this project. I am grateful to the center's administrative director Krista Birch, to two consecutive Williams College faculty directors, Jana Sawicki and Gage McWeeny, and to a number of the Oakley fellows who gave me important comments and suggestions. I would also like to acknowledge the help that I received from the knowledgeable and extremely efficient staff at the Institute Archives and Special Collections, Massachusetts Institute of Technology, where the Thomas S. Kuhn Papers are kept.

Two people deserve a special mention. Evan Pence, my research assistant, contributed enormously to the clarity and readability of the primary texts, and helped me complete and update all of Kuhn's references. He also made many excellent philosophical points and suggestions, most of which I was, regrettably, unable to address in this volume; but I am still thinking about them. Mane Hajdin read carefully and constructively all of the editorial contributions to the volume, and gave me many insightful and useful suggestions. His support was vital for the completion of this work, as it always is.

Bojana Mladenović

More than twenty years have passed since Thomas S. Kuhn's untimely death. The book that made him famous, *The Structure of Scientific Revolutions*,[1] has achieved the status of a classic: it is indispensable reading for every well-educated person. It is increasingly recognized that Kuhn was not only one of the most important philosophers of science but also one of the most important thinkers of the twentieth century, whose influence reached and, in some cases, thoroughly transformed a number of academic fields.[2] To be sure, some of Kuhn's views are still as controversial as they were in 1962, when *Structure* burst upon an audience still steeped in logical empiricism, but his philosophy is now much better understood than before, and its complexity and nuance are much more appreciated.

This is in no small measure due to Kuhn's own sustained efforts to explain and defend the central claims of *Structure*. In time, however, he became persuaded that further clarifications—however careful—would not do; he came to think that his philosophy of science needed to be revised to some extent, and that it also needed to be situated within a larger, reworked philosophical framework. He published a series of papers in which he presented an overview of the new direction that his philosophy had taken.[3] This work was to culminate in a new magnum opus, a book that was his main project for more than a decade; sadly, he did not live to complete it.

This volume finally brings to the public eye all of the drafted chapters of this eagerly awaited book, provisionally entitled *The Plurality of Worlds: An Evolutionary Theory of Scientific Development*. This manuscript is preceded by two related texts, not previously published in English: Kuhn's paper "Scientific Knowledge as Historical Product" and his Shearman Memorial Lectures, "The Presence of Past Science." The volume also includes two abstracts, one for the Shearman Lectures and the other for *Plurality*. Although they are editorial creations, the abstracts use Kuhn's own formulations whenever possible. They show, at a glance, the areas of thematic overlap between the two works. In addition, the abstract for *Plurality* sketches the main issues with which the unwritten parts of the book were to be concerned, insofar as these could be responsibly reconstructed.

This introduction to the volume consists of three parts. Part I presents the history of the three manuscripts, their relation to one another, and their state. Part II, intended mostly for readers not thoroughly familiar with Kuhn's post-*Structure* philosophical preoccupations and development, provides that information and context, and sketches the contours of the book *Plurality* was intended to be. It is, in a way, a road map through the complicated, often overlapping, and fundamentally unfinished primary material.[4] Part III of the introduction offers concluding remarks on the nature and contents of this volume.

I. The Contents of This Volume

Sources

In working on this volume, I relied on a number of sources. Although I do not discuss here all of Kuhn's previously published texts, or the rich secondary literature on Kuhn, these works did give my editorial work a necessary background. Some of the articles that Kuhn published in the late 1980s and the 1990s were especially helpful, since that is where the philosophical project of *The Plurality of Worlds* begins to take shape.[5] Even more important was Kuhn's foreshadowing, in the drafted chapters of the manuscript, of what was to come later in the book. In addition, Kuhn left a rich archive of unpublished texts of various kinds, most of which are kept at the Institute Archives and Special Collections, Massachusetts Institute of Technology. The most important among them, for the purpose of reconstructing Kuhn's unfinished book, are the Thalheimer Lectures,[6] Kuhn's class notes and handouts for his MIT graduate seminars, in which he often discussed his book in progress,[7] and his correspondence with colleagues, especially his exchange of letters with Quentin Skinner in the wake of the Shearman Lectures.[8]

An important source that I relied on when reconstructing *Plurality* is not publicly available, however: the unrevised notes that Kuhn left for each projected chapter of the book.[9] For the most part, these notes are brief and suggestive rather than detailed and explicit; nonetheless, I found them very useful in producing the abstract for *Plurality*.[10] Jehane Kuhn, Kuhn's widow and literary executor, gave me a copy of transcribed conversations among Kuhn, James Conant, and John Haugeland, in which she occasionally participated.[11] The conversations took place in Kuhn's home, June 7–9, 1996, in five working sessions, totaling about seven hours. Kuhn wanted the tapes of the conversations destroyed, and he never meant the transcripts to be publicly available.[12] Out of respect for Kuhn's wishes, I did not use these transcripts as a source of information about his philosophical views, but only to reconstruct the history of his work on the manuscripts published in this volume.

None of the sources provides anything approximating a rough draft of the unwritten parts of *Plurality*. Rather, they give us a sense of Kuhn's general philosophical direction, with very clearly stated reasons, here and there, against a particular misunderstanding of his views, or against a rival philosophical position that might be mistaken for Kuhn's own. Thus, the available sources throw only a partial, ambient light on the project of *Plurality*, which Kuhn was still thinking through in June 1996. No one can know now what would have been the final, detailed account of his view had he had the time to articulate it fully; but the overall contours of his position can be sketched, and at least some details filled in.

Primary Texts

"Scientific Knowledge as Historical Product" and Kuhn's Shearman Memorial Lectures, "The Presence of Past Science," are both philosophically important on their own and significant as milestones in the development of the ideas central to Kuhn's unfinished book. Arranged chronologically, the three texts show Kuhn's philosophical trajectory from the 1980s until his death, in 1996.

"Scientific Knowledge as Historical Product" was drafted and revised multiple times between 1981 and 1988. Various versions of it were given as invited lectures.[13] In his first Shearman lecture, Kuhn notes that "Scientific Knowledge as Historical Product" is "to appear in *Synthèse*" (meaning *Revue de Synthèse*, a French journal of history and philosophy of science), but it did not appear there.[14] The last version, included in this volume, was given as a lecture in Tokyo in 1986 and subsequently published in *Shisō* in Japanese translation.[15] It offers the best available account of Kuhn's analysis of the origins and commitments of the traditional epistemology of science, the problems that plagued it, and the ways in which Kuhn's developmental understanding of science avoids these problems. Although there is no significant textual overlap between this paper and the opening chapter of *The Plurality of Worlds*, the two texts share the same title and perform the same function of justifying Kuhn's developmental, historically sensitive, practice-oriented philosophy of science. I tend to think of this paper, then, as a proto–chapter 1 of *Plurality*.

"The Presence of Past Science" is a series of three Shearman Memorial Lectures that Kuhn gave at University College London in November 1987. The lectures explore Kuhn's developmental-historical approach to science and begin to articulate the philosophical consequences of adopting it. Two other lecture series preceded them: the Notre Dame Lectures, "The Nature of Conceptual Change," delivered at the University of Notre Dame in November 1980, which appear to be lost;[16] and the Thalheimer Lectures, "Scientific Development and Lexical Change," presented at Johns Hopkins University in November 1984.[17] The Shearman Lectures are the latest complete version of Kuhn's mature philosophy, and the best available—if imperfect—guide to what his book aimed to accomplish: they sketch the whole philosophical landscape that the projected book was to cover. The last lecture is particularly important in giving us a sense of what would have been the content of part III and of the epilogue of *Plurality*, had Kuhn lived to write these parts of the book.

Kuhn did not publish the Shearman Lectures, nor any other lectures that he gave in the late 1980s and early 1990s. He treated them as more or less successful drafts of his book. He did, however, revise and polish the manuscript of the Shearman Lectures, and he shared it with a number of his colleagues, friends, and students; it is still in semiclandestine circulation in some philosophical circles.[18] The Shearman Lectures thus became a major unpublished textual source for appreciation of Kuhn's later philosophy. Two splendid articles—the first by Ian Hacking and the second by Jed Buchwald and George Smith[19]—analyze and discuss the Shearman Lectures in philosophically stimulating ways, rich in nuance and detail; a full understanding of these articles, as well as of Kuhn's published response to Hacking,[20] requires familiarity with Kuhn's original text. So, since the Shearman Lectures are by now widely discussed but not generally accessible, and since the book that was to supersede them was not completed, Kuhn's literary executors and the University of Chicago Press decided that this important text should be included in this volume despite Kuhn's original intention not to publish it.[21]

The centerpiece of this volume is, of course, Kuhn's unfinished book, published here under the working title at the time of Kuhn's death: *The Plurality of Worlds: An Evolutionary Theory of Scientific Development*. Had Kuhn lived to complete the book, he would probably have given it a different title. The original working title seems to have been *Words and Worlds: An Evolutionary View of Scientific Development*. This is the title Kuhn proposed in his successful application for a 1989 National Science Foundation grant in history and philosophy of science.[22] It is not clear why Kuhn abandoned this title, which adequately announces the intended content, nor why he did not return to it when he became concerned that his *The Plurality of Worlds* might be confused with David Lewis's *On the Plurality of Worlds* and mistakenly assumed to be, like Lewis's book, about modal logic.[23] Kuhn expressed this concern to Jehane Kuhn, who told me of it in a private communication in 2017. Kuhn's wish to find a new title for his book is also documented in his transcribed conversations with James Conant, John Haugeland, and, in this segment of the conversation, Jehane Kuhn.[24] Speaking about the title, Kuhn said that it should include *worlds*, or *plurality*, but he decided to leave the final decision to Jehane, who did not change the title.

Kuhn's plan for the book was an ambitious one, and the work on it took a very long time.[25] It was to open with acknowledgments and a preface, followed by three substantive parts, each consisting of three chapters: part I, "The Problem"; part II, "A World of Kinds"; and part III, "Reconstructing the World." An epilogue was to be added, and an appendix was to conclude the work. Unfortunately, complete drafts exist of only part I (chapters 1–3) and chapters 4 and 5 of part II; the draft of chapter 6 is unfinished. Kuhn left sparse notes for part III and the epilogue, but no actual text; the preface and appendix are also missing.

Part I is polished, clearly close to the intended final version. It motivates the project of the book as a whole, and outlines the planned chapters ahead. Its focus is on the nature and philosophical significance of historical study of science, vividly introduced through detailed case studies of Aristotle's, Volta's, and Planck's works. Kuhn used these three case studies to show how exactly history of science must confront incommensurability in order to produce understanding and to pose the important philosophical questions that the last part of the book was designated to address. Although there is a considerable textual overlap between the first Shearman lecture and chapter 2 of *Plurality*, the overall differences between the two works, separated by less than a decade, are also considerable, and very important in revealing the trajectory of Kuhn's thinking and the development of his mature philosophical position. The second Shearman lecture, for example, discusses incommensurability between past and present science, and sketches the contours of theories of meaning and knowledge that would allow us to make sense of historical understanding despite incommensurability. Insofar as this lecture gestures toward an empirically based account of language learning and concept acquisition, it is the germ from which part II of the book was developed; but the actual text and philosophical methodology differ greatly.

In fact, part II—in contrast to part I—will likely be a great surprise to readers familiar with Kuhn's published writings. Kuhn seems to be searching here for a naturalistic foundation of his prospective theory of meaning, which should, in turn, ground his revised idea of incommensurability. He aims to use the results of scientific research in cognitive and developmental psychology as a basis for his theory of meaning and understanding across incommensurably different lexical structures and practices. This important project is advanced, but not completed, however. I suppose that the final version of part II would have updated and compressed the relevant results of scientific research, and then highlighted their philosophical significance, thereby preparing the ground for the philosophically most interesting—but unwritten—last segment of the book.

Part III was to twist together the historical view of conceptual change, on display in part I, and scientific accounts of concept acquisition presented in part II, in order to explain both incommensurability and our ability to understand and communicate in spite of it. *Plurality* treats incommensurability as ubiquitous across cultures, languages, historical periods, and various social groups; scientific communities divided by incommensurability are but a special—albeit very special—case. Kuhn aimed to explain both the way in which science shares universal patterns of concept acquisition and structuring of lexicons, and the way in which lexical change in science differs from lexical change in natural languages. General philosophical questions about meaning, understanding, belief, justification, truth, knowledge, rationality, and reality were all raised by Kuhn's project, and he meant to address them in part III. The main goal was to develop theories of meaning and of knowledge that would take incommensurability as their starting point, and find room for, first, a robust notion of the world that science investigates, second, for the rationality of belief change, and finally, for the idea that scientific development is progressive.

The epilogue was to return to the question of the proper relationship between history and philosophy of science, which preoccupied Kuhn since *Structure*, and which magnetized both his critics' and his admirers' attention. In his early work, Kuhn passionately argued against presentist (or anachronistic) approaches to history of science, which he saw as characteristic of both logical empiricism and Popperian falsificationism.[26] He was convinced in *Structure* and in his 1977 book of essays *The Essential Tension* that philosophy of science must reject presentist case studies, and rely on responsible, detailed historical work that recovers context, concepts, problems, and intentions of past scientific communities. However, in the late 1980s, Kuhn came to think that presentist historiography has its own irreplaceable function, which he was to explain and discuss in the epilogue to *Plurality*. Fortunately, this central idea for the epilogue is very clearly presented in the last Shearman lecture.[27]

Finally, the appendix was to offer a detailed comparison between the views presented in *Structure*, which remained the source of Kuhn's central philosophical ideas as well as of the main problems that preoccupied him until the end of his life, and *Plurality*, which was to be his final word on these issues.[28] The continuities and differences between the two works were to be highlighted and explained. Insofar as we can accurately reconstruct Kuhn's last book, we can also imagine what the substance of the comparative appendix would have been.

But to reconstruct Kuhn's unfinished book in sufficient detail is not an easy task. We are obliged to rely on various texts—published and unpublished—outside the manuscript itself. They were written over more than a decade, and it is not always clear which of the ideas that Kuhn explored in this period he intended to fully articulate and defend, and which he would have rejected in the final version of his book.

Insofar as part III can be reconstructed, then, I tried to do so in the abstract that I created for *Plurality*. This still leaves the reader with only a skeletal representation of the centerpiece of Kuhn's book. It is thus important to bear in mind that the publication of the manuscript does not, by itself, fully represent Kuhn's ambitious philosophical project. Its proper appreciation requires interpretive and imaginative efforts different in kind from the efforts that were needed to understand the unfamiliar landscape of *Structure* at the time of its publication; but now as then, the effort will pay.

II. A Guide to Kuhn's Unfinished Project

From *Structure* to *Plurality*

Reacting against philosophical approaches to science dominant in 1962, when *Structure* was published, Kuhn insisted that science should be seen as a historically developing set of traditions, through which knowledge changes and grows. Scientific change is neither uniform nor strictly cumulative; rather, it exhibits a two-phase pattern. Periods of normal science, marked by consensus within the scientific community on all fundamental matters, produce coherent, cumulatively progressive results. When this consensus breaks down under the pressure of accumulated anomalies, the scientific community enters a period of extraordinary science, marked by

competition between proponents of rival, incompatible frame-
works for doing science, which Kuhn in *Structure* called *paradigms*.[29]
These rivals are incommensurable, and the eventual choice of one
among them is not forced by either logic or paradigm-neutral empirical evidence. Scientific revolutions are thus disruptive episodes
of fundamental reconfigurations, through which scientific knowledge develops in a noncumulative way.

The reception of *Structure* was not what Kuhn was hoping for.
In his view, both his critics and would-be followers seriously misunderstood the book.[30] He was read as a radical relativist, whose
views cannot explain scientific change as due to good reasons and
evidence, but only as a result of rhetorical, institutional, or political
power of the side that ultimately won. Thus, it was argued, Kuhn
cannot see science as the paradigmatically rational enterprise that
gets us progressively closer to the truth about the world.[31] Moreover, Kuhn's startling claims—that "when paradigms change, the
world itself changes with them" and "though the world does not
change with a change of paradigm, the scientist afterward works
in a different world"[32]—inspired charges of idealism and constructivism. Kuhn rejected such characterizations of his view, while
maintaining that some of his paradoxical-sounding claims are actually correct. For the rest of his extremely productive career, he
was to return to *Structure* in the hope of making its claims both
understandable and plausible.

His post-*Structure* philosophical work can be seen as developing
through at least two relatively distinct periods.[33] The first period
starts with the 1969 postscript to the second edition of *Structure*
and ends in the early 1980s.[34] Kuhn was then responding to numerous mischaracterizations of his book with clarifications, explanations, and new arguments, but without dramatic revisions.
He argued that incommensurability does not imply impossibility
of communication or comparison, and that scientific choice is
not primarily driven by social and political power. Insisting on
the communal nature of scientific inquiry, Kuhn highlighted the
importance of rigorous, formative scientific training and of the

shared values that guide all scientific research and evaluation.[35] He began to stress that scientific reasoning and practice cannot be separated, and must be understood as products of a scientific group that, through its expert judgment, choice, and practice, constitutes science as a rational inquiry into various aspects of the world. Nonetheless, characterizations of his position as hospitable to radical relativism, irrationalism, and social constructivism persisted; Kuhn's consistent rejection of such characterizations was still rarely taken seriously in this period.

In the mid-1980s, Kuhn's work entered a new phase, which I refer to, interchangeably, as "Kuhn's mature philosophy" or "the late Kuhn." All three texts collected in this volume are from this period, during which Kuhn undertook more radical revisions of *Structure* and considerably widened his philosophical concerns. He came to distinguish among the different perspectives from which working scientists, historians, and philosophers ask their questions about science. This led to a more nuanced, qualified, and precise understanding of incommensurability as ubiquitous but local, and of scientific change as revolutionary only when seen from a great historical distance. Most importantly, Kuhn concluded that his philosophy of science needed a general theory of meaning, a full-blooded epistemology, and a novel take on the debate between scientific realism and constructivism. His main task was then to reconfigure these fields in such a way that his view of scientific development as involving incommensurability between historically distant theories and practices would both make sense and not jeopardize the general perception, which Kuhn wholeheartedly shared, of science as rational and progressive.

Historicism

It is typical for Kuhn to open a philosophical text by stressing the importance of history as its necessary starting point. The first sentence of *Structure*—"History, if viewed as a repository for more than anecdote or chronology, could produce a decisive transformation in the image of science by which we are now possessed"[36]—could easily be seen as the motto for all of his subsequent work. For Kuhn, philosophical reflection on science needs to be grounded in accurate description of actual scientific practice and of its meandering history, since without a proper understanding of how science works and changes, philosophy of science cannot explain either its successes or its failures.

Kuhn's historicism was in sharp opposition to philosophical projects of logical empiricists and Popperian falsificationists, which were both primarily normative rather than descriptive, and fundamentally uninterested in the history of science.[37] Their goal was to develop and justify a set of methodological rules that reliably lead to increased scientific knowledge, and thus explain progress in science. This tradition had no great use for meticulous historical research but rather relied on simplified, decontextualized descriptions of some episodes in history of science, perceived as crucially important from the present-day point of view. Kuhn thought that this normative-methodological philosophical project and anachronistic, presentist historiography reinforce each other, and jointly create a distorted image of science. This image Kuhn sought to replace with a properly diachronic and descriptively accurate image of his own.

Kuhn's own approach to history was hermeneutic; that is, internalist and contextual. Hermeneutic historical narratives strive for explanatory success through maximal consistency, completeness, and avoidance of anachronistic explanatory categories and distinctions. Passages that seem incomprehensible or obviously false to a present-day reader should be valued as the essential puzzles for a historian to solve. For the late Kuhn, hermeneutic historiography is a kind of retrospective ethnography, which aims to understand concepts, beliefs, and practices that, to the historian, initially appear alien and often absurd.[38] Serious historical narratives may focus on great scientists, important experiments, or momentous discoveries, but they always provide historical context and background. In that sense, they are always about whole scientific communities, whose concepts and beliefs the historian tries to recover. He must re-create in his narrative the web of commonly shared assumptions and beliefs, typical argumentative strategies, nodes of disagreement, and the intended audience of scientific writings. Most importantly, the historian needs to master the structured lexicon of past scientific communities, a lexicon that is typically incommensurable with his own. Historical understanding is thus akin to learning a long-lost language, with only partial and often misleading connections to the language of current science. The goal is to create a narrative within which past beliefs and choices can be seen as reasonable and plausible, rather than irrational, mistaken, or absurd.

Although *Structure* influenced sociology of science and inspired careful historical research in that field,[39] Kuhn was strongly opposed to the sociological explanatory categories that structured these historical narratives.[40] Sociologists of knowledge represented scientists as primarily engaged in political or social power struggles, and argued that scientific choice must be explained as determined by personal idiosyncrasies, ambitions, and, especially, political interests. Kuhn took this to imply a skeptical conclusion about the cognitive authority of science and rejected such historical narratives as incapable of accounting for the importance of empirical observation and experiment in driving scientific change. Sociologists of knowledge, in Kuhn's view, do not pay enough attention to scientists' self-understanding as explorers of nature, and thus they cannot explain either what scientists do, or why they do it. His own hermeneutic historiography privileges *cognitive* explanatory categories and is strictly internalist and intentionalist.

Kuhn's understanding of philosophical uses of history evolved throughout his career. His last period shows three important developments. First, he gave even greater prominence in his philosophical texts to actual historical case studies. For example, in the Shearman Lectures and in *The Plurality of Worlds*, three case studies—of Aristotle, Volta, and Planck—are foregrounded and presented in much more detail than the historical examples in

Structure. This method of exposition is, for Kuhn, also a method of thinking: his view of science and of incommensurability are not illustrated by case studies, but rather emerge from them. A deeper involvement with specific historical narratives in his mature period allowed Kuhn to locate the sites of incommensurability with much greater precision than in his early works, and to then raise general philosophical questions concerning meaning, rationality, ontology, truth, and progress on firmer grounds than before.

Second (and surprisingly for some), in the last decade of his life, Kuhn recognized that we need presentist historiographical narratives as much as we need the hermeneutic ones. Hermeneutic historiography remains uniquely suitable as the starting point for philosophical reflections on science, as is shown through the case studies that open both the Shearman Lectures and *Plurality*. However, the *motivation* for philosophical reflection on science as a supremely rational and progressive quest for knowledge can come only through presentist narratives.[41] Presentist narratives project present-day scientific concepts, questions, and problems onto the past, and trace precursors as well as obstacles to our own ways of doing science. This does not lead to an understanding of past scientific communities on their own terms—quite the contrary— but it does help us feel *connected* to them. Moreover, the late Kuhn concluded that his analysis of scientific development needs a dose of presentism in order to account for scientific developments as truly *progressive*; they can be seen as such only from the present point of view. Although incompatible with the hermeneutic approach, presentist historiography needs to be done parallel with it, for—as the third Shearman lecture suggests and as the epilogue to *Plurality* was to expand upon—it is only through presentist historiography that the past can be seen as *our* past. The late Kuhn's acceptance of the multiplicity of legitimate kinds and uses of history replaced his earlier belief that only one kind of historiography has real value for philosophers of science.

Finally, the late Kuhn refined his articulation of the *status* that history has in his philosophical project. His historicism was too often mistaken for an empirical theory, in which historical data are supposed to provide straightforward evidence for his cyclical model of scientific change. He took great pains to distance himself from this interpretation, highlighting instead that the main value of his historiographical work was to help him develop a historical *perspective* on science. A historical perspective is a way of seeing, a sensibility, developed through a deep involvement with internalist hermeneutic historiography, but argumentatively unencumbered by a historian's concern to produce explanatory narratives of particular events. Once acquired, this perspective naturally gives shape to the questions that philosophers need to ask about science, and it also suggests solutions to some of the problems that plagued the reception of *Structure*.

The most significant among them was the problem of explaining the periods of extraordinary science as periods during which rational discourse continues to play a crucial role in scientific work. *Structure* highlighted numerous conceptual, methodological, and practical incommensurabilities among rival paradigms, and asserted that their proponents often talk past one another, relying on different standards of cogent reasons and of empirical evidence. To Kuhn's critics, this image of extraordinary science seemed to collapse into a radical, almost self-refuting relativism. Kuhn was read as saying that incommensurability between rival paradigms is *complete*. Without any shared conceptual, methodological, or evaluative grounds, the ultimate choice of one of the rival paradigms cannot be rational; worse still, in the absence of a common language, the disagreements between the proponents of rival paradigms cannot even be stated. Of course, Kuhn never meant to defend such a position, but he did realize that his description of extraordinary science could be misleading. He concluded that, in his early work, he did not sufficiently distinguish between the perspective of contemporaries in the midst of a fundamental disagreement, and the perspective of a historian writing many centuries after the events he was trying to describe and understand.

From the perspective of the actual historical actors, all trained in the same way, immersed in the same practice, and facing the difficulties and anomalies that they all recognize as such, it is always possible to understand what an opponent is saying, the late Kuhn realized. At any given time, all members of a scientific community have much in common. Revolutions *appear* to be swift, decisive, and complete changes only from a considerable historical distance, because incommensurabilities between the rival ways of doing science grow over time. From the standpoint of the scientists themselves—as well as for a historian who focuses only on a short, crucial period of extraordinary science—changes cannot but be described as incremental and partial, always justified with an appeal to the shared beliefs, methods, and values that are not, at that moment, called into question.

In his mature period, Kuhn preferred to discuss the *evolution* of scientific knowledge through a process he metaphorically linked to speciation. He no longer identified scientific revolutions as periods in which a new paradigm replaces an old one, but rather, he saw them as periods in which an old way of doing science effectively *splinters* into a number of newly formed specialties: the old domain of phenomena becomes divided among different new disciplines, as do the basic methods, problems, and solutions that survive the revolution. Looked at this way, revolutions should be depicted as the speciation-event nodes on the phylogenetic tree; the resulting specialties are the branches that shoot off from such nodes. The role of incommensurability in Kuhn's new model of scientific change is extended as well: it now obtains not only between the old and new lexical structures and practices, but also among the new specialties themselves. Each will study its own domain of phenomena, with very small areas of overlap with the others; each will develop what Kuhn came to call a complete structured lexicon, incommensurable with the structured lexicons of the other disciplines.[42]

Naturalism

For philosophers who think of historicism and naturalism as polar opposites, the structure of *The Plurality of Worlds* is likely to be puzzling, to say the least. Kuhn's well-known historicism, on such splendid display in part I, seems to vanish in part II, to be replaced with detailed reports of scientific experiments in cognitive and developmental psychology. Although Kuhn never used the term *naturalism* to characterize his philosophical project, his reliance on results of scientific research does make him a naturalist of sorts.[43] His suggestion that part III will return "to the themes of part I, for which part II attempted a *foundation*," requires an explanation, however.[44] First, it is not clear how empirical results of psychological experiments could provide a foundation for answers to the philosophical problems raised by incommensurability. Second, Kuhn consistently and explicitly rejected epistemic foundationalism. Both as a historian and as a philosopher, he would always start in the middle of things, consider concepts, beliefs, and practices as already in place, and then he would ask what motivates and what justifies a particular *change* to any of them. What kind of foundation could such a situated epistemology need, and whatever for?

Despite the way he described the task of part II of *Plurality*, Kuhn never thought of the research reported in it as providing *epistemic* foundations for his philosophical project. His epistemology is not in search of certainty—it is not even especially interested in distinguishing between belief and knowledge. Rather, by *foundations* Kuhn meant the *starting point* of human cognitive development, and the innate neurological basis that will be activated in all subsequent concept acquisition. All human beings share this biological basis for cognition, and concept acquisition in all of us follows the same developmental trajectory. Kuhn does turn to scientific research to discover what these innate capacities are, how flexible they are, and how they develop from infancy to potentially multilingual adulthood. This does make him a naturalist in one of the many senses in which that label is used among philosophers, but it is important to note that his naturalism is neither reductionistic nor scientistic. It does not intend to replace philosophical questions about meaning and knowledge with a summary of scientific research on early concept formation. Rather, it seeks to ground and constrain the questions that philosophers can reasonably ask about conceptual change. Had he lived to revise part II and write part III, it would have been evident that he turned to science in exactly the same spirit, and for the same kind of reason, that he initially turned to history.

To see this, recall that Kuhn argued that, in order to understand science, we must understand its history; a changing, evolving practice cannot be properly understood if its diachronic nature is not appreciated. Hermeneutic internal historiography provided the best means of doing so, and Kuhn took it for his starting point ever since *Structure*. This historical approach revealed incommensurabilities among differently structured scientific lexicons. To understand what makes such different lexicons possible and effective, and to what extent we communicate across incommensurability and how, Kuhn needed a descriptively accurate account of our capacities for acquiring, systematizing, using, and changing our concepts. The best source of that information was not history but psychology, and so he turned to the research on categorical perception, cutting-edge at the time, to gain reliable information about biological and developmental aspects of human conceptual capacities. Arguably, had he lived longer, he would have enriched his understanding of lexical structures with relevant research from evolutionary biology and linguistics, especially sociolinguistics. Although scientific research in the relevant fields has considerably developed since the 1990s, when Kuhn was working on *The Plurality of Worlds*, the general structure of his philosophical project is not undermined: it is meant to incorporate *whatever* the best scientific research delivers on human conceptual development and capacities.

Thus, it is not merely the case that Kuhn's historicism and his naturalism are not in tension with each other. In fact, they are but two different ways of respecting the same reasonable requirement: that the phenomena on which a philosopher reflects first be accurately described. History of science describes scientific development and change, recovering in its narratives past problems, lexicons, canons of reasoning, and other aspects of scientific theory and practice. Scientific research in evolutionary biology, cognitive and developmental psychology, and linguistics describes the capacities and processes involved in creating lexical structures. Kuhn's historicism and his naturalism thus both answer to the descriptive demands of his philosophical project, and constrain the questions that can reasonably be asked about incommensurability, understanding, and, most importantly for Kuhn, the practice of science.

Concepts, Kinds, and Structured Lexicons

The first task of *The Plurality of Worlds* was to develop a theory of meaning, capable of explaining meaning change, the ubiquity of incommensurability, and human ability to overcome the barriers it presents to communication and understanding. Given Kuhn's general philosophical orientation, it is not surprising that his theory was to be structurally different from other available theories of meaning. Instead of asking, as traditional theorists do, What is meaning? Kuhn raised several interrelated *developmental* questions: How are concepts acquired? How are the meanings of words learned? Why do the meanings of some words change over time? What is a conceptual change, and how does it happen? In other words, Kuhn was searching for a dynamic, developmental, and descriptive theory of concept *acquisition* and meaning *change*.[45]

In his mature philosophy, Kuhn already thought of incommensurability as a local phenomenon, reaching global proportions only when seen from a great historical distance. In his last works, he explicitly extended this point from the history of the natural sciences, where he first noticed it, to a general view about human languages, which are frequently locally incommensurable with one another. The late Kuhn argues that the key sites of incommensurability, whether among natural languages or among specialized scientific theories, are to be found in clusters of interrelated *kind terms*. In *Plurality*, Kuhn sought to describe two consecutive developmental paths that need to be traced if we are to understand the process through which highly specialized scientific kind terms are articulated. The first is the path of individual human cognitive development, from birth to bilingualism; the second is the path of communal development of lexical structures, from kind terms in natural languages to abstract technical terms of mature science.

The biological basis of human capacity to categorize objects into kinds is present from birth in its rudimentary form. In part II of *Plurality*, Kuhn discusses the empirical evidence supporting the view that human infants are born with specific neurological structures that function as modules for acquiring concepts. First in development is a protoconcept from which, at a later stage, the child will acquire the concepts of *object*, *space*, and *time*; after that, concepts of *cause*, *self*, and *other* will follow. The prelinguistic structures for classifying and reidentifying individuals are generally modifiable through experience, so it is perhaps best to think of them as *inborn flexible capacities* for the acquisition of full-blooded concepts.

The innate capacities for learning a language are very broad; they can be activated by any human language, none of which is easier or more natural to acquire than the others. However, these capacities can be activated only through repeated interactions with competent speakers, who support and correct the learner throughout a trial-and-error process of mastering the language's structured lexicon. By *structured lexicon* Kuhn means a framework constituted by sets of projectable kind terms, typically hierarchically organized. Mastering a kind term requires mastering other kind terms in the same taxonomic cluster, as well as mastering contrasting clusters within the same lexical structure. Empirical research on concept acquisition supported Kuhn's Wittgensteinian rejection of the traditional account of concepts as defined by necessary and sufficient conditions.[46] In his view, experiments in the field of categorical perception suggest that recognition of an object as an object of a particular kind does not require knowledge of features shared by all members of the kind, contrary to what the traditional theory of concepts says. First, for most natural kinds, there simply are no such universally shared features. Second, and even more important, recognition of an object as being of a certain kind depends on *noninferential perception* of relevant similarities and differences, learned on particular examples and entrenched through agreement and corrections by other accomplished speakers.

All members of a linguistic community use the same categories, and they cluster objects in the same way, even if they differ here and there in how they describe the kinds that they use. Structured lexicons are thus essentially collective, but their taxonomies are not universal. Certain similarities and differences are seen as salient in one language, while they may be unimportant in another. Natural languages thus develop structured lexicons that often turn out to be incommensurable with one another. Since the perception of an individual object is inevitably a perception of it *as* an object of a particular kind, and since natural languages have, to some extent, different kind terms and different lexical structures, to master a language is also to become socialized into a particular culture and to see the world through the lens of its natural and social taxonomy. Kuhn always stresses that language and world are learned together: the world is, as it were, unveiled by the acquisition of a language through the mastery of kind terms. This gives a community its sense of what sorts of things exist in the world, and how they behave in it. In Kuhn's words, a structured lexicon gives an ontology to its users, and it greatly restricts what the community members' beliefs could be.[47] This is not always obvious to the language speakers themselves. Although any lexicon provides only a particular, contingent, changeable, and fully replaceable lens through which the world can be seen and interacted with, the categories of one's first language tend to be experienced—at least initially—as natural and inevitable. As language users, we are not always aware of the degree to which our lexicons actively structure—that is, both enable and limit—our understanding of the world.

Incommensurable ontologies stand in the way of perfectly accurate translations. Kuhn insists that this is not at all an insurmountable barrier to either understanding or communication. The
inborn cognitive modules that enable us, as infants, to learn our
first language continue to provide a basis for mastering new lexical structures: we can all be proficient in more than one language,
and if we are, we are likely to sometimes vividly experience difficulties of translation, without being in any way deprived of full
understanding. Bilinguals thus have a cognitive advantage over
monolinguals: it is easier for them to realize that the natural world
does not impose any particular lexical structure on human beings.[48] In their practical lives, however, bilinguals need to navigate
a much more complex social world. They must be constantly aware
in which linguistic community they are currently participating:
their thoughts, speech, and actions are all shaped by the lexical
structure in which they think and live, and some aspects of their
being in the world—especially the social, communicative world—
cannot be simply transplanted from one language to another.[49]

For Kuhn, then, bilingualism is a cognitively demanding but reliable bridge across incommensurability. Whether we are thinking
about very different natural languages spoken in different parts of
the world, or about now dead languages of times past, or indeed
about technical languages of various kinds of specialists, understanding through bilingualism is always possible. Acknowledging
Wittgenstein, Kuhn concludes part I of *Plurality* by suggesting that
if something is to count as a human language, then it can, in principle, be understood by other human beings. Our neurological equipment for novel concept acquisition provides a slim but serviceable
basis for mastering new lexicons; moreover, understanding across
incommensurability is aided by our common human biology and
the shared environment of our planet. This suggests that some
natural-kind terms will, as a matter of fact—that is, not as a matter
of necessity—exist in every human language.

To refine his account of structured lexicons, Kuhn began to
develop in *Plurality* a taxonomy of kind terms. He distinguished,

first, between natural and artefactual kind terms. Natural-kind terms in ordinary language aim to sort observable objects found in the world by similarity and difference; paradigmatic examples are names of species, such as *ducks* or *swans*.[50] Natural-kind terms are projectable: to master them is simultaneously to accept some claims about regularities of the behavior of their referents. Natural-kind terms cannot overlap in their referents, unless they are related as species to genus; Kuhn called this the *no-overlap principle*. This principle imposes the need to restructure the lexicon when the community comes across an anomalous individual that seems to belong to two different kinds. For example, a warm-blooded furry animal, with a duck-like beak and webbed feet, that lays eggs but then feeds the young with milk produced by its mammary glands, was understandably causing a considerable confusion in the minds of eighteenth-century European naturalists. Was the specimen a mammal, a bird, a reptile, or a hoax?[51] Our present-day ability to confidently classify platypuses as monotremes is due to the taxonomic revolution brought by the Darwinian theory of evolution. Kuhn points out that scientific experts will increasingly emerge as the group responsible for making taxonomic decisions about newly encountered anomalous phenomena, and thus they will sometimes create revisions, or deep restructuring, of the community's lexicon.

In contrast to natural kinds, artefactual objects—paradigmatic cases being everyday human-made objects, especially tools—are not grouped into kinds by similarity and difference of their observable features but exclusively by their function. Moreover, not all artefacts are observable. Some, such as *goodness*, or *money*, are what Kuhn calls unobservable *mental constructs*.[52] They are learned through their relation to other mental constructs within a practice. Some of them are what Kuhn calls *singletons*, in contrast to taxonomic kind terms. Taxonomic kind terms are mastered within a hierarchy, and learned together with their appropriate contrast sets (for example, to learn how to recognize swans, a child must learn that swans are not ducks, but that they are both fowl). The meaning of a taxonomic kind term is thus bound up with the meanings of the other kind terms in the same set; none has meaning independent of the others. Singletons are not situated on any taxonomic tree, and they do not have contrast sets: they are sui generis. Kuhn sometimes says that both taxonomic kinds and singletons are governed the by no-overlap principle, but there are passages in his notes in which he seems to doubt whether the no-overlap principle really applies to all singletons. At the time of his death, he was still struggling with the proper characterization of singletons. Although they play important roles in natural languages, Kuhn was especially interested in them because of their vital role in mature science. For example, *mass* and *force*, the key terms of modern physics, are singletons: they are neither a genus nor a species on a taxonomic tree, nor does either term have a contrast class. To explain the role of singletons in science, and their enormous importance as the primary loci of incommensurability between historically distant scientific communities, Kuhn offered an account of how structured lexicons of mature science develop from natural languages.[53]

In tracing this developmental path, Kuhn noted that the main purpose of taxonomies in natural languages is to classify objects detectable through the senses, such as plants, animals, or visible celestial bodies. Early science begins with inquiry into the nature of such objects; this results sometimes in reclassification of some of them, sometimes in refinement or sharpening of classificatory boundaries, and sometimes in the creation of new taxa. In the process, early science also creates new artefactual kinds: objects to be used as tools and instruments in the inquiry, and abstract concepts for explanatory and predictive purposes.[54] Lexical structures of mature science develop from all of these resources and achievements of early science. Although mature science continues to discover previously unknown natural kinds (such as new species, materials, or celestial bodies) and to adjust existing taxonomies to accommodate them, it becomes progressively more concerned with artefactual rather than with natural kinds.[55] Structured lexicons cease to be limited to the classification of pretheoretically individuated objects and instead give the central place to newly forged abstract terms, such as *mass* and *force* in physics, or *gene* in biology. Many of these terms are interrelated singletons, introduced together with one or more universal generalizations, often in a mathematized form. For example, it is impossible to learn the meaning of Newtonian *force* without knowing Newton's second law of motion, $F = ma$. The importance of singletons for Kuhn's mature philosophy is enormous, because these terms are the ones primarily involved in revolutionary conceptual change: Newtonian *mass* is not Einsteinian *mass*, although the two terms are not mere homonyms, either, since the later concept developed from the earlier one, restructuring it completely within the new theoretical framework.

This led Kuhn to believe that, in contrast to members of natural kinds, scientific singletons are never observable.[56] This, however, should not be understood as Kuhn's return to the logical empiricists' distinction between observable and theoretical terms. Kuhn clearly wanted to avoid *that* distinction, with its implied givenness of observation, since he thought of scientific observation as possible only through an already available conceptual structure—even though that structure can, and often does, change. Unfortunately, Kuhn did not live to fully think through the important similarities and differences between his view of unobservable referents of singletons and the logical empiricists' concept of observation terms.

Possible Worlds of Science

The last chapter of *The Plurality of Worlds* was to answer two questions that preoccupied Kuhn since *Structure*: What could a real world be? What, if not correspondence to the real, gives truth its constitutive role in science?[57] Although the extant texts do not give enough information to answer either of these questions on Kuhn's behalf with confidence and in sufficient detail, I will sketch the general direction that I think he wanted to take. I will start with his question about the world, leaving the question of truth for the next section.

When he wrote in *Structure* that "though the world does not change with a change of paradigm, the scientist afterward works in a different world," Kuhn was fully aware of the paradoxical nature of his claim. "Nevertheless," he immediately added, "I am convinced that we must learn to make sense of statements that at least resemble these."[58] In subsequent papers and lectures, as well as in *Plurality*, he tried to provide a solution to what came to be known as his *world-change problem*: how to explain the crucial role that the world plays in scientific inquiry, while preserving his insight that the world in which scientists work actually changes after a revolution. It is noticeable that the scope of the world-change claim is wider for the late Kuhn than it was for the author of *Structure*. *Structure* discusses world changes only in the aftermath of scientific revolutions; the late Kuhn thinks that world changes occur *whenever* a significant conceptual change happens, especially when it involves the restructuring of old kinds. For example, after a radical conceptual change in political, cultural, or aesthetic discourse, communities live in a new and different world. Science is special not because its development involves dramatic conceptual reconfigurations, for that is a widespread phenomenon; it is special because of the very stringent local criteria that urge, constrain, and justify conceptual change.

It might be helpful to bear in mind that, in trying to explain what he meant by his world-change claim, Kuhn did not want to endorse either straightforward scientific realism or straightforward constructivism. In a similar vein, he tried to develop an account of natural kinds that would avoid both traditional metaphysical realism and traditional nominalism. Of course, there are many differences among those who would consider themselves realists about science in general, and about natural kinds in particular; the same is true of those who see themselves as continuing the constructivist or the nominalist tradition. In rejecting both sides in both debates, Kuhn is certainly rejecting the uncomplicated versions of these positions. Whether he really rejects all forms of realism, constructivism, and nominalism will depend on how these positions are stated *precisely*. I will not attempt to do this here, however. Since my interest is to outline the contours of Kuhn's position as far as his texts give me a warrant to do so, I will focus on that task. The reader may very well conclude that it is possible to classify Kuhn's position as a peculiar member of one of these families of views, despite his resistance to such classifications.

It is very clear that Kuhn was neither a traditional realist nor a traditional constructivist. It is useful to think of traditional realism with respect to science as having three components. The *ontological component* asserts the existence of the world as a mind-independent reality: it is as it is, independently of our language, categories, needs, or desires. The *semantic component* states that scientific theories aim at truth, where *truth* is understood as correspondence, or isomorphism, between our beliefs and the world. A realist then holds that all scientific statements are true or false in virtue of what the world is like. This, in turn, requires that all nonlogical terms in scientific theories (including kind terms and singletons) are capable of referring to real-world objects and structures. Finally, the *epistemic component* of scientific realism states that mind-independent reality is, at least in part, knowable by us, and that science provides the most reliable way to acquire that knowledge. A scientific theory is *better* than its rival if it is *closer to the truth* than its rival is. Consequently, when a realist thinks that science is progressive, she thinks that its progress consists in the fact that scientific theories of a later date are closer to the truth about the mind-independent world than the earlier theories.

In a similar vein, traditional realists about natural kinds think of these as mind-independent groupings that structure the world prior to, and independently of, human language, needs, or interests. They see the natural-kind terms as aiming to mirror these independently existing groupings, to faithfully capture the real similarities and differences among things in the world. Our concepts justifiably change only when we learn more about the way that the world is really structured. For example, a realist thinks that, although this was not always known, it was always the case that the Sun is a star and not a planet, that dolphins are mammals, and that water is H_2O. One of the functions of science is to discover *real* natural kinds and their taxonomy; we will then revise our lexicons accordingly.

A constructivist will reject or reinterpret all components of realism, starting with a denial that there is anything that we could coherently call a mind-independent world. Everything that we can say is expressed in our categories, and guided by our expectations and needs. We cannot step outside our concepts to verify whether they adequately represent the world. Thus, we cannot know what the world is *really* like. Forever confined to our representations, we cannot even compare scientific theories to the world, nor scientific statements with facts, nor distinguish between referring and nonreferring terms; we can only compare a theory with other theories, one set of statements with another, one way of categorizing what we call *the world* with another way. All our categories are shaped by our expectations, needs, and desires; some systems of categories serve our purposes better than others, and so we prefer them. Scientific theories aim to satisfy some of our needs: for example, our need for accurate predictions, for successful manipulation of our environment, for a coherent system of beliefs, or for explanations that make sense to us; those theories that satisfy our needs better are better scientific theories overall. It is easy to see that constructivism is hospitable to a traditional nominalist view about natural kinds: a nominalist does not believe than there are any *natural* kinds in the realist's sense. All groupings of things, all kinds into which they are sorted, are human inventions, driven by human needs and interests.[59]

It is clear that Kuhn wants to reject all of these well-known views. Against traditional constructivists and nominalists, he believes that the world is mind-independent and that it imposes constraints on what a useful lexicon can be. The very fact that we encounter some objects that seem to violate the no-overlap principle, and thus force us to restructure our preexisting taxonomy in order to classify them, suggests that some taxonomic solutions are better than others, not just in terms of our preferences but also in terms of their adequacy. It is not the case that *just any* grouping would do as well as any other, however much our interests and desires may favor it. In a similar vein, Kuhn thought that it makes no sense to think of the world as constructed or created by human beings. In his mature period, he was unambiguous on the subject:

> First, the world is not invented or constructed. The creatures to whom this responsibility is imputed, in fact, find the world already in place, its rudiments at birth and its increasingly full actuality during their educational socialization, a socialization in which examples of the way the world is play an essential part. The world, furthermore, has been experientially given, in part to the new inhabitants directly, and in part indirectly, by inheritance, embodying the experience of their forebears. As such, it is entirely solid: not in the least respectful of an observer's wishes and desires; quite capable of providing decisive evidence against invented hypotheses which fail to match its behavior. Creatures born into it must take it as they find it.[60]

Against the traditional natural-kinds realist, however, Kuhn rejects the idea that the world is already fully structured and divided into kinds, waiting for our most precise lexicon to reflect the divisions in nature. No lexicon *simply* mirrors nature. The categories that we use to orient ourselves in the world are of our own making; the ones that we use now are not the only ones that could enable such an orientation. The world could be differently described and its elements differently categorized, as we can see by examining different human languages. Although the world constrains our lexical choices, it does not favor a *single* one. Multiple, mutually incommensurable lexicons could each give us knowledge about the world. Realist metaphors of correspondence and isomorphism suggest a one-to-one match between our kind terms and the groupings in the world, and thus they are not apt for conveying a plurality of possible ways in which the world could accurately and usefully be described and organized for action.

To better understand Kuhn's unusual way of thinking about the world and natural kinds, we must start where he does: with the claim that human experience in general, various specific human practices, and science in particular all require *some* categorization into kinds. Kind terms are thus essential for both ordinary language and science.[61] Part II of *Plurality* was to show that human brains are preprogrammed, as it were, to see the world as sorted into kinds. We could not experience the world as consisting of randomly distributed properties, without any objects; nor could we experience it as containing a variety of objects without any significant similarities or differences among them. At one time, Kuhn seems to have thought of inborn capacities for categorical perception as being akin to a Kantian a priori contribution to human cognition. Upon reflection, however, he concluded that the distinction between the a priori and the a posteriori aspects of experience is untenable. In *Plurality*, he sought to characterize categorical perception as a contingent and flexible result of an equally contingent evolutionary process, which tends to favor those features and capacities that turn out to be helpful in survival. In this respect, human lexicons have the same evolutionary basis as the capacities of other animals to categorize the world, but our specifically human languages greatly increase our ability to see the world as a world of kinds.[62]

It is in this sense that a structured lexicon provides an ontology to which our words apply. Within it, our words do refer to objects in the world. Some kinds emerge *as natural* within our lexicon. Kuhn says that this makes our natural-kind terms transparent: we see the world through them.[63] Kind terms enable and guide our interactions with the world, including our observations of it. We do discover various properties of members of natural kinds by direct observation, but, Kuhn insists, *which* properties are in fact observed will depend on the available kind set; and the selection and structure of kind sets are both deeply influenced, although not fully constituted by, human interest and purpose. However, Kuhn argues, since there is a plurality of possible lexicons, there is also a plurality of ways to individuate kinds and sort objects into them. We can learn to *see* in new ways by mastering new lexical structures.[64]

This is, in fact, what scientists do. Early scientists asked questions in their natural languages, using the available kind terms for particular species, materials, or celestial bodies: they eventually came up with answers that effectively restructured parts of the everyday lexicon in which their original inquiry was formulated. This process of repeated lexical revisions led to mature science, expressible only by means of highly technical terms. In Kuhn's view, every lexicon makes certain questions, beliefs, and practices *possible*. "What one is committed to by a lexicon is not therefore a world but a set of possible worlds, worlds which share natural kinds and thus share an ontology. Discovering the actual world among the members of that set is what the members of scientific communities undertake to do," he writes in the conclusion of the second Shearman lecture.[65] But the set of possibilities allowed by a lexicon is limited. To explain certain anomalous phenomena, and even to ask some novel questions, communities sometimes need to restructure a part of their scientific lexicon to "gain access to worlds that were inaccessible before."[66] One way to characterize normal science for the late Kuhn is to see it as the quest for the actual world among the possible worlds that the shared lexicon allows. Revolutions are then gradual but ultimately radical openings of new sets of possible worlds, unimaginable for ancestral scientific communities. In that sense, it could be said that different scientific communities work in different worlds that they themselves both discover and create. As Kuhn said, we must learn to make sense of statements like this one.

Truth

Kuhn frequently stated that the correspondence theory of truth needs to be replaced by a theory that can make better sense of scientific practice. He hoped to present a new theory of truth in chapter 9 of *The Plurality of Worlds*, for which he left only sparse notes. Before trying to imagine what his theory of truth would have been, we must ask why he thought that a new theory was needed in the first place.

The correspondence theory is probably the most natural, and consequently the most widely held theory of truth; it has also attracted considerable criticism throughout the history of philosophy. Kuhn, however, did not seem to have any one of the standard objections in mind when he rejected the correspondence theory. Rather, he was dissatisfied with the two ways in which truth—understood as the correspondence theory understands it—figures in a number of general philosophies of science.

First, Kuhn found deeply problematic the widespread view that truth is the goal of inquiry—that science aims to provide true theories about the world, and progresses by ever closer approximations to that goal. In his view, we cannot make sense of the claim that truth pulls the development of science forward, since the ultimate truth is epistemically inaccessible from the standpoint of scientific communities faced with difficult choices. For the same reason, a philosopher of science cannot explain why scientists accepted, rejected, or modified particular beliefs by reference to truth as the ultimate goal of science. Rather, the explanation has to be given in terms of reasons and evidence available to those who made the choice. Approximation to truth would not do, either, since there could be approximations that are mutually incompatible, but equidistant from truth.[67] So, Kuhn concluded, neither correspondence truth nor approximate truth could constitute the goal of science, or explain scientific belief change and progress.

His other reason for rejecting the correspondence theory of truth was that he saw it as implicated in presentist historiographical narratives favored by logical empiricists. Philosophers who asked whether past scientific beliefs were true were, in Kuhn's view, very much misguided about the proper way to understand the past. In his internalist approach to history, the question of truth or falsity of past scientific beliefs arose only as a question about how past scientific communities distinguished between true and false beliefs. The question of whether their beliefs were *really* true actually makes no sense, according to Kuhn, unless we trivialize it by translating it into the anachronistic question of whether the beliefs in question are true by *our* lights.

Kuhn was equally dissatisfied with all other theories of truth that he considered. The coherence theory of truth cannot give empirical observation the special status that, in Kuhn's view, it needs to have; this is especially problematic when describing scientific debates, in which empirical evidence plays a large, although not always a decisive, role. Kuhn also rejected what he thought of as two versions of the pragmatist theory of truth: truth as warranted assertability, and truth as the ideal end of inquiry. He argued that truth cannot be analyzed as warranted assertability, because this analysis violates the logic of truth statements. Two people, holding logically incompatible beliefs, may each have a warrant, but at most one of them could be saying something true—a situation that the warranted-assertability theory of truth cannot explain.[68] Finally, Kuhn thought that truth defined as the ideal consensus at the end of the inquiry would not help at all in explaining what we do when we *currently* take some beliefs to be true and others to be false.[69] The end of inquiry is epistemically inaccessible, and thus not available to explain the beliefs that were actually held, and the choices that were actually made, by scientific communities.

Kuhn did not live to formulate a new theory of truth, capable of playing a role in his philosophy of science. This is perhaps not surprising, since a philosophical theory of truth is not what his philosophy in fact needed. There are two different contexts in which he needed to *use* the concept of truth, but neither of them required a fully fleshed-out general theory of what truth *is*.

The first is the communicative context of scientific inquirers, members of the same scientific community. Kuhn insists that the logic of communication requires that every discourse has the means of prohibiting contradictions and of marking some beliefs as true and others as false. Evaluation of beliefs as true or false is, for

Kuhn, simply a condition of communication. The criteria that govern this kind of evaluation are shared across the community, but different epistemic communities may develop different criteria for truth. In that sense, the criteria are internal and local. No community will be able to distinguish true beliefs from those that, after the most rigorous scrutiny, *appear* to be true. At the time when a belief is considered, there are no markers of its truth in addition to the markers of its rationality and plausibility, in light of the best evidential and inferential reasons available to the community in question. From an external point of view, the distinction between the rationality of a belief and its truth is both obvious and important, but it is not epistemically available to those who are facing the choice of accepting, revising, or rejecting a belief.

The second context in which Kuhn reflected on the use of the concept of truth is that of the history of science. A historian looks back to a scientific lexicon that long fell out of use; the beliefs, methods, and practices associated with it are alien to the scientific community active in the historian's own time. A statement made in the new lexicon is often a different statement than a statement made in the old lexicon. Most interesting statements of past science elude straightforward translation: Kuhn believes that what they say is *ineffable* in the later lexicons. Since past scientific beliefs cannot be simply restated in a modern vocabulary, they cannot be simply evaluated as true or false, either. The historian's task is thus to explain why past beliefs were reasonable and plausible in their own epistemic context. To use Kuhn's favorite example, a historian discovers that Aristotle had excellent reasons, both conceptual and evidential, for thinking that there is no void. To understand what he really meant, we cannot simply translate his claim into our language; we have to understand it within *his* lexical structure, within the system of beliefs framed by his assumptions; and in that context, it is not merely true but tautological that there is no void. While denying that he was a relativist about truth, Kuhn accepted that he was a relativist about effability: meaning of words and sentences is context-relative, and the larger epistemic context itself, with its lexical structure, assumptions, beliefs, and practices, cannot be evaluated as true or false. It can be evaluated in other ways, Kuhn points out—for example, for its effectiveness in serving the goals for which it was put to use—but that is an interest-driven evaluation, not an evaluation of truth or falsity.[70]

It is clear that, in making these points, Kuhn focuses on rationality, justification, and plausibility of beliefs. These need to be distinguished from truth, however. The most reasonable and the most scrutinized beliefs of an epistemic community may turn out to be false; and the beliefs it justifiably deems irrational or unsupported by reason and evidence, may be true. Although Kuhn does not deny the importance of the distinction between truth, on the one side, and rationality, justifiability, and plausibility, on the other, and although he does stress that scientists are interested in the truth and not only in the rationality of their beliefs, his philosophy makes *no use* of this distinction. When he examines scientific discourse from the imagined standpoint of its participants, Kuhn notes that the distinction between truth and rationality is not epistemically available for evaluation of beliefs. When he thinks about science as a hermeneutic historian, he is not even interested in the truth of past beliefs, but only in their reasonableness by local lights. He thus seems, on the one hand, to acknowledge the distinction between rationality and truth, but then to undermine it, on the other. This is, I believe, because he recognizes the importance of the distinction only in an abstract philosophical context. As a historian of science, or as an active participant in his own scientific community, Kuhn actually sees *no point* in drawing it. He may very well have been justified in this indifference; but then, it seems, he should also have seen that his philosophy of science has no need for a novel theory of truth.

And in fact, Kuhn does not seem at all to be engaged in the endeavor of formulating and defending such a theory. Nowhere in his writings, published or unpublished, do we find an inquiry into what truth *is*, or what truth-makers are. Instead, Kuhn is interested in *the pragmatics* of discourse that evaluates statements as true or false, and the specific epistemic requirements and resources that a community has at its disposal to distinguish between the claims it deems true and those it deems false. Kuhn wants primarily to understand how the predicates *true* and *false* are *used* within a communicative practice.[71] It seems that he wanted an account of what scientists in a particular community mean, and what they do, when they say that a statement is true or that it is false. But if these were his questions about truth, then there was nothing abstract and general that he should have offered. He did not need a philosophical theory of truth to replace the correspondence theory. He only needed to reiterate his rejection of philosophical reliance on presentist historiography, as well as his rejection of the idea that truth—however analyzed—is the goal of science. After stating that every community needs to sort beliefs into true or false, and that the logic of truth talk must respect the principle of noncontradiction, he could have left to a historian or ethnographer the search for specific answers about specific communities and their criteria for evaluating beliefs as true or false. This made Kuhn uneasy, but it shouldn't have. In everyday life as well as in scientific communication, *truth* is readily understood and unproblematic.

The difficulties that Kuhn had in articulating a plausible view of the role that truth plays in science are probably one of the main reasons why part III of *Plurality* remains unfinished. I suspect that these difficulties were due to an incongruity between the ways in which questions about the nature of truth are traditionally posed in philosophy, and Kuhn's historicist and pragmatist way of thinking about scientific communities. Perhaps, rather than try to work within a framework alien to his way of thinking, Kuhn should have simply set aside the question about the nature of truth, in the same spirit in which he set aside radical skeptical challenges.

III. Concluding Remarks

The vivid impression that Kuhn was first and foremost a philosopher of scientific practice may be a bit dimmed by his last writings, focused as they are on language, meaning, and structured lexicons. However, it would be a mistake to conclude that, in his mature period, he came to think of science primarily in terms of its linguistic and theoretical aspects. The absence of explicit discussion of practice in Kuhn's mature philosophy is due to two factors. First, he was satisfied with what he had said about it in his earlier works, and felt no need to improve or expand on that. The focus on lexical structures in his final period was due to his growing sense that incommensurability—the central concept of his philosophy—has not yet been analyzed and explained in sufficient detail. Second, a careful reading of his last works will show that practice remains central to his view of science. It is no longer foregrounded but, rather, woven into the developmental questions that he asks about language acquisition and use; it remains the lens through which he sees all philosophical problems that preoccupy him. For Kuhn, no less than for Wittgenstein, questions about meaning tend to be recast as questions about learning and use. Similarly, to learn a language is to learn how to be in the world: what to perceive, how to organize and report perceptions, what to say, and how to act. This is as true of scientists as it is of everyone else, but *the scientists'* ways of being in the world are mediated by highly complex intellectual lexical structures that Kuhn sought to understand. Throughout, he continued to see practice as crucial, and scientific expertise as consisting in largely tacit knowledge of how to see problems, how to classify phenomena, and how to search for a solution.

The philosophical questions about meaning, reality, truth, and knowledge that preoccupied Kuhn in his last years only apparently take the abstract form characteristic of traditional epistemology and metaphysics. His thinking actually always starts from, and returns to, scientific practice. For example, his lasting insight about incommensurability does not concern only meanings of kind terms or difficulties of translating between differently structured lexicons. As he repeatedly pointed out in his post-*Structure* writings, incommensurability between languages is not an insurmountable obstacle to either contemporary communication or retrospective understanding. Robust incommensurability in Kuhn's philosophy concerns *doing*, not saying or understanding. Scientific communities divided by the incommensurability of their problems, lexicons, and evaluative standards can still *make sense* of each other's projects, but they cannot collaborate on them—they cannot *do* science together. Had he lived to complete *The Plurality of Worlds*, he would have stressed, in part III, the priority of scientific practice over theory, and the relative independence of the former from the latter.[72]

At the end, a crucial question faces both the editor and the reader of this volume: Can an unfinished work be a successful one? The straightforward answer seems obvious, and negative: Kuhn did not live to complete *Plurality*, and what is published here is not what he wanted to see in print. But the success of his last work need not be measured only by its distance from the intended goal; we can also measure it by the distance from its starting point in *Structure*'s revolutionary ideas. Throughout his intellectually intense and prolific life, Kuhn modified, developed, and restructured these ideas, adding nuance and expanding their applicability. If we take a developmental perspective on Kuhn's last writings, we will see the texts published here as but a moment in his mature rethinking of a young man's valuable insights. We will also see the philosophical method that he developed through this process fully at work in his last writings. The mature Kuhn seamlessly moves between particular, detailed case studies and synoptic philosophical considerations, bringing the latter back to bear on his—thus refined— understanding of scientific practice and its development. Kuhn's dynamic method of perennially searching, restructuring, focusing, and expanding would have never ended in a definite conclusion or the final resting of his case; but that, I think, is what success in philosophy might look like.

EDITOR'S NOTE

All primary texts are published with only minimal editorial inter-
ventions. I completed and updated Kuhn's references, corrected
misspellings, and here and there added an obviously missing word.
All editorial additions, including references in footnotes, are in
square brackets. Kuhn used the same brackets in some of the quotations, to mark *his* additions, and I did not change any of these.

At the end of each of the first five chapters of *The Plurality of
Worlds*, Kuhn noted the date of the last revision. I left these dates
in place.

Kuhn's notes to his texts are in footnotes; endnotes are editorial additions.

I created abstracts for the Shearman Lectures and for the last
draft of *Plurality*, using Kuhn's own formulations whenever possible. The two abstracts will show, at a glance, the areas of thematic overlap between the two works. In addition, the abstract for
Plurality sketches the main issues with which the unwritten parts
of the book were to be concerned, insofar as these could be reconstructed from the drafted chapters and from the (publicly unavailable) notes that Kuhn left for each chapter. I am grateful to the
University of Chicago Press and to the Kuhn family for allowing
me access to these notes.

Scientific Knowledge
as Historical Product

: : :

Thomas S. Kuhn

I shall open this lecture with a brief autobiographical statement of intent.[a] Nearly a quarter century ago I was one of a group of emerging scholars who, almost simultaneously and virtually independently, attacked the long-dominant tradition in empiricist philosophy of science.[b] What was called science by the tradition bore so little resemblance to what scientists do, we proclaimed, that the relevance of its conclusions to what they produced was doubtful. Did the tradition, we rhetorically asked, actually deal with scientific knowledge at all? Our answer was a resounding no (in retrospect I think our stridency was excessive), and our evidence was drawn mostly from history of science. That evidence we used also to launch what we took to be a more adequate approach.

To most of us, however, history seemed primarily a convenient source of data about actual science, data which might in large part have been gathered without exhuming the past. I, for example, once wrote: "Actual experience in the practice of science would probably be a more effective bridge [across the gap between philosophers of science and actual science] than the study of its history. Sociology of science . . . might do as well."[1] By the same token, though we all viewed science as an essentially human enterprise,

---

1. Thomas S. Kuhn, "The Relations between the History and the Philosophy of Science," chap. 1 in *The Essential Tension: Selected Studies in Scientific Tradition and Change* (Chicago: University of Chicago Press, 1977). The quotation is on p. 13 and the lecture from which it is taken was delivered in 1968.

none of us thought to emphasize that it must ipso facto be essentially historical. Looking back, I think we missed the primary source of our views. The most central respects in which the new philosophy of science departed from the old were less a response to the facts of history than to the perspective it provided. Where the tradition had concerned itself with science as a static body of knowledge, our concern was necessarily with a dynamic, developmental process. Science became for us a sort of knowledge factor, and that shift proved more important than the data it revealed in generating the new philosophy of science.

Today, I shall attempt an overview of that change of perspective, trying to display the form taken within the developmental approach to philosophy of science by problems—especially the problem of theory evaluation—central to the tradition which preceded it. That goal will call for comparisons. To provide a basis for them, I begin with a brief epitome of the tradition which the still-developing historical approach to philosophy of science aims to displace.

The tradition's central feature has been known as foundationalism, and most of its other presently relevant aspects follow from that one. Like so much else in modern philosophy, foundationalism originated with modern science during the seventeenth century. Bacon and Descartes are its first major proponents. Both proclaimed the impotence and unreliability of the knowledge claims of their predecessors; both blamed these shortcomings on inadequate method, observational as well as intellectual; and both believed that circumstances demanded a fresh start. "There was but one course left, therefore," Bacon wrote at the start of *The Great Instauration*, "to try the whole thing anew upon a better plan, and to commence a total reconstruction of sciences, arts, and all human knowledge, raised upon the proper foundation." "The mind itself," he continued, must "be from the very outset not left to take its own course, but guided at every step; and the business be done as if by machinery."[2] And Descartes resolved in his *Discourse*, "to arrange my thoughts in order, beginning with things the

2. *The Works of Francis Bacon*, ed. James Spedding, Robert Leslie Ellis, and Douglas Denon Heath, vol. 8, *Translations of the Philosophical Works* (New York: Hugh and Houghton, 1869), 18, 60–61. The first passage is from *The Great Instauration*, the second from *The New Organon*.

simplest and easiest to know, so that I may then ascend . . . step by step, to the knowledge of the more complex," accepting at each point "nothing as true . . . additional to what had presented itself to my mind so clearly and so distinctly that I could have no occasion for doubting it."[3] Their tones were different, but their notions of what the method must achieve were the same. Certain knowledge was to be constructed step by indubitable step upon an indubitable foundation.

About the nature both of the foundations and of the ascent from them, Bacon and Descartes disagreed. Put far too simply, the foundations were believed by Bacon to be empirical, by Descartes to be innate; correspondingly, the ascent from the foundation was for Bacon inductive, for Descartes mathematical and deductive. In philosophy of science the tradition that descends from their work has adopted elements from each. Ordinarily, it has followed Bacon in insisting that, with the possible exceptions of logic and mathematics, the foundations of scientific knowledge must be empirical, the critically scrutinized testimony of the senses. But it has also ordinarily followed Descartes in looking to mathematical proof for a model of the step-by-step connections between those foundations and the conclusions that they support. Each choice has been responsible for the emergence of additional characteristics of the tradition, including certain of its characteristic problems.

Consider first the empirical foundation. If it were to provide a basis for certain knowledge, then the observations and experiments from which it was constituted must themselves be certain, accessible to and binding for all normal human observers. Observations with such universal authority must be independent of all cultural and personal idiosyncrasy. More particularly, they must be *purely* observational, embodiable in reports that are *purely* descriptive. All resort, direct or indirect, to prior beliefs must be eliminated from those reports. They must describe unadorned, uninterpreted sensation.

---

3 . René Descartes, *Discourse on Method*, in *Descartes' Philosophical Writings*, ed. and trans. Norman Kemp Smith (London: Macmillan, 1952), 129. The lines are from part II of the *Discourse*, and the order of the two fragments is here reversed.

Conceptions of what a pure observation, or a pure observation report, would be have varied considerably over the last three centuries. But characteristically they have been taken, either literally or as an ideal, to be systematically constructible from elementary sensory elements—colors, shapes, smells, and so on—elements that would be identified in the same way by all people with normal sensory apparatus. "Red there," accompanied by pointing, is then a simple or atomic sensory report; "red triangle there," a complex or molecular one. Reports of the presence and behavior of middle-sized physical objects—whether falling apples or expanding metals—were to be compounded in the same manner and thus acquire the same objectivity as the elementary sensations of which they were made up. Any pure observation report could in principle be restated in terms of these elementary sensory givens. Though neither this nor any other program with the same objective—operationalism, verificationism, and so on—has ever been successfully developed, insistence that all observation reports be compoundable from indubitable elements has remained a central characteristic of the tradition. Representative attempts run from Locke's "simple ideas of sense" through Russell's "knowledge by acquaintance" and Wittgenstein's "elementary propositions."[c] The continued frustration of these and other efforts has been a central difficulty for the tradition.

Several other characteristics of the tradition as well as its other central difficulty result from the choice of the deductive mathematical model to connect the concrete empirical foundations of knowledge with the general conclusions those foundations support. The aim of both Bacon and Descartes had been a truth-certifying method of discovery. The intent of their methods was, thus, constructive, the certainty of the foundations being transmitted bottom-up to each new floor. But observations are always of the singular, the concrete, and deduction can, despite Descartes's hopes, lead only from the more general to the more particular, from high-level axioms and postulates, for example, downward to particular theorems. Only after hypothetical laws or theories have been arrived at, by whatever means, can deductive methods be applied to them. Those methods work from the top down, educing not new generalizations but the consequences of those already at hand. No method that aims at the certainty of mathematics can generate discoveries. This aim of the tradition was soon abandoned.

The deductive model was not, however, abandoned with it. What deduction can do is generate testable conclusions from generalizations already at hand, and the result for the tradition has been the introduction of an increasingly categoric distinction between the so-called contexts of discovery and of justification.[d] The first is concerned with the route by which scientists arrive at generalizations. Once hope for a constructive method had been given up, discovery was relegated by the tradition to psychologists and sociologists. Only justification, the evaluation of proposed laws and theories, remained the proper concern of philosophy of science. As the logical empiricists, in particular, have emphasized, laws and theories may come into being in many ways: accidents or personal idiosyncrasy often play a role; the special concerns and training of the investigator always do. But it is not their manner of generation that renders the resulting innovations contributions to knowledge. The same processes could as well have led to egregious error, and whether they have in fact done so can be determined only by one or another form of testing, validation, confirmation. These processes lie within the context of justification; it is when they are deployed that deductive methodology functions; and it is they alone which the tradition has taken to be of philosophical concern.

Having narrowed the tradition's focus to problems of evaluation, the mathematical model continued to give it special shape. If deductive methodology was to apply, the knowledge claims to be evaluated had necessarily to be embodied in a set of timeless statements. Scientific knowledge thus came to be viewed as a collection of propositions—statements, that is, whose truth or falsity is independent of the time, the circumstances, and the language of their utterance. The philosopher's problem was, correspondingly, to specify rational techniques for determining which of the propositions in the collection—which of the generalizations in a scientific text, for example—were true and which false. Proposed solutions took the form of logical relationships that were to provide criteria of acceptability.

Some of those criteria were internal to the set of propositions that embodied the knowledge claims. Of these, consistency was the most obvious and nearly standard; simplicity, a notion notoriously more difficult to make precise, was often a second; and there were others besides. Still more important was a second group of partially external criteria. No observation statements, no propositions embodying empirical data available at the time, were to conflict with any of the set of knowledge claims or with their deductive consequences. Closeness of fit between observation statements, on the one hand, and laws and theories, on the other, was then one condition of acceptability. Another was scope, the range and variety of observation statements that could be matched to consequences of laws and theories. Again, there were others besides.

Both justificationism and what I shall barbarously call propositionalism survived a further alteration of the tradition, the abandonment in many circles of the insistence that method must result in certain knowledge. However many tests a law or theory passes, it may still fail the next one that confronts it. Satisfying test criteria, like those above, can only make a theory probable, not certain. Much effort within the tradition has therefore been devoted to the development of probabilistic techniques for evaluating theories. But none of these attempts altered the presently important characteristics of the tradition. They were consequences of the insistence on a mathematical model, not on that model's particular form.

Two such characteristics have already been noted. First, what was to be evaluated was a static body of propositions, the cognitive content of science or some part of science at a given time. Second, only considerations specifiable in terms of relationships between propositions could be relevant to an evaluation's outcome. Two more characteristics follow, the first sometimes known as methodological solipsism.[e] Like mathematical proof, the outcome of an evaluation was necessarily coercive, determinable by and binding upon any rational individual. An evaluation that required judgment and thus allowed rational individuals to differ was seen as tainted by subjectivity. In principle, therefore, any rational individual could substitute for any other in objective evaluation, and only a single individual was required. Science thus became a one-person game. Not the nature of science but the limited power of human beings necessitated the participation of others, whether over time or at a time.

Finally, all evaluations, to the extent that they were rigorous, proved also to be holistic. Because any evaluative procedure involves a number of propositions, a failure necessarily reflects on them all. Usually there are plausible reasons for attributing such failure to a small subset of the propositions involved, but the attribution cannot be certain, only plausible. What can be tested with the certainty at which the tradition aimed is therefore never an individual knowledge claim but only a body of them, and the size of that body of knowledge claims has proved remarkably difficult to restrict. Under the name of the Duhem-Quine thesis, this characteristic of the tradition has emerged in this century as the second prominent barrier to the realization of its founders' hopes, the first being the previously noted difficulties in implementing the concept of an indubitable empirical base.[f]

An empirical foundationalism and a deductive justificationalism have, in sum, been the two primary goals of the main tradition in philosophy of science. With justificationalism, furthermore, has come solipsism, propositionalism, and an unwanted holism. How, let me now begin to ask, do these aspects of the tradition look from the viewpoint of the still developing historical approach to philosophy of science? What is the effect upon them of what I earlier called the altered perspective from which the historically oriented view the sciences? Historians necessarily see science as a continuing process, one lacking any starting point at which knowledge acquisition might begin from scratch. All narratives of scientific development start in midstream, with the scientific process already underway. Whenever they open, their protagonists already possess what they take to be a relatively complete body of knowledge and belief[s] about nature. Though they recognize that some things remain to be known, those are not for the most part the things subsequent scientists will turn out to discover.

Under these circumstances the historian who wishes to recount the development of one or another set of laws and theories has two tasks, each with significant implications for philosophy of science. First, he or she must discover and explain how these older doctrines (often most strange and apparently implausible) could ever have been accepted by intelligent people as the basis for a long-enduring tradition of scientific practice. Second, the historian must seek to understand how and why the status of those beliefs changed, what led to their displacement by another set, the frontiers of research shifting with the change. For the historian, in short, unlike the traditional philosopher of science, the advance of science is marked less by the conquest of ignorance than by the transition from one body of knowledge claims to a different, though overlapping, set. An account of that transition must therefore first display the integrity of the older set and then examine its displacement.

In this lecture I shall be primarily concerned with the implications for philosophy of the second of these tasks, with the consequences, that is, of viewing the emergence of new knowledge as a displacement of old rather than as an advance into previously empty territory. But I shall close by talking briefly about the historian's prior task, that of discovering and reconstituting the integrity of an out-of-date scientific tradition. In the long run, this reconstitutive task will, I expect, prove the more consequential of the two.

Considering the entry of new ideas into the sciences, my concern is primarily with aspects of justification, arguably the central problem of philosophy of science. Within the tradition its basic form has been, "Why should one believe a given body of knowledge claims?" From the newer, developmental viewpoint the question is instead, "Why should one shift from one body of knowledge claims to another?" The old criteria—consistency, simplicity, scope, closeness of fit, and the like—continue to function when the question is put in this way, but their function is now comparative or relative, not absolute, as it had been before. With respect to closeness of fit, for example, one need no longer ask, "Does the body 'X' of scientific laws and theories fit observation reports well enough to be accepted?" Because fit is never perfect, that formulation inevitably raised the further question, "How well is 'well enough'?," and no one has even suggested what a generally acceptable solution to that question might look like. Within the developmental approach, on the other hand, one asks only, "Does the body 'X' of laws and theories fit observation reports better than body 'Y'?" An important source of equivocation is then eliminated. Questions of simplicity, scope, and so forth transform in the same way and with the same result. Elimination of the need to set an apparently arbitrary standard for acceptability is a first consequence of seeing scientific development, not as the acquisition of new knowledge where none existed before, but as the replacement of one body of knowledge claims by an overlapping but different one.

A second significant difference is closely related. As evaluation becomes comparative, considerations once thought relevant only to the context of discovery become critically important to the context of justification as well. To understand a scientific discovery or the invention of a new scientific theory one must first find out what the members of the relevant scientific specialty knew, or thought they knew, before the discovery or invention was made. In addition, one must determine what part, if any, of that earlier body of knowledge claims needed to be set aside and replaced as the innovation was accepted. While evaluation was considered absolute, considerations like these belonged to the context of discovery alone, but comparative evaluation places them in the context of justification as well. It is only the altered knowledge claims consequent upon innovation that require justification, not the body of claims that remains unchanged, common to the new view and the old. Those beliefs, whatever their ultimate fate, are simply not at risk in the choice between bodies of knowledge to which they are common. Justification thus requires knowledge of the body of beliefs accepted by scientists immediately before the innovation to be evaluated was made.

The point is not that discovery and justification are the same process, but that many of the considerations relevant to the first prove central to the second as well. Indeed, in the first stages of either, the overlap is usually so great that even the processes themselves cannot be told apart. A discovery for which there is as yet no justification is ipso facto no discovery at all. Though further testing is often required after a discovery has been made, there is no discovery to be tested until there is already evidence in its favor. Explore the historical record as closely as they may, historians regularly find it impossible to say when, in the continuing development of scientific knowledge, discovery ceases and justification begins, which experiment or bit of conceptual analysis belongs to each.

Implicit in what has just been said is still another difference between the static and the developmental approaches to philosophy of science. As the latter conceives justification, only the innovations—the knowledge claims that distinguish the new body of belief[s] from the old—are placed at risk. The problem of holism posed by the Quine-Duhem thesis thus appears to have been solved. But what has, in fact, occurred is something more fundamental, not so much a resolution as a dissolution. Holism was a by-product of the way in which the problem of justification was posed by the static tradition, and it has no equivalent within the developmental approach. Once justification becomes comparative, one can no longer be pressed to holism. Though the logic of the Duhem-Quine thesis remains impeccable, it no longer has a bearing on justification. Always starting in midstream, a philosopher of the developmental school can seek good reasons only for *change* of belief. Though the beliefs shared by the positions being compared are vital to the arguments of both sides, their justification has no bearing on the choice a scientist must make between those positions, and it is on that choice that the cognitive status of science depends.

Asked what justifies the acceptance of current science as the basis for further scientific practice, the developmentalist can only reply with another question: Can any rational alternative to acceptance even be conceived? The closest approach, one supposes, would be provided by tracing the historical path to the present body of belief[s] and justifying each of the individual decisions made along the route. But not all relevant past decisions are accessible. In any case, discovering an example of irrational choice among those that are [accessible] would neither render the body of current belief irrational nor permit the clock to be turned back and the alternate road traced. When the point at issue concerns an entire body of belief[s], justification is simply not at issue.

A scientist *must* accept a great many of the current knowledge claims of his community, for they are constitutive of the community's practice, of the form of life, that is, of an enduring tribe. To refuse to accept them would be to decline membership in that tribe and thus to refuse the practice of science. Though there is a great deal about which members of an individual tribe—physicists or chemists, for example—may disagree, those very disagreements are made possible, recognizable, and discussable only by the far larger body of beliefs that members share, beliefs that unite them as members of a single tribe. That much of their body of belief[s] is simply among the givens of a given time. Discovering those givens is what the historian must do to recapture the integrity of an older mode of thought. If more time were now available to discover how the historian does so, phrases like "constitutive of the community's practice" would come to seem less like mere bits of incantation.

To say all this, however, is simply to say that the traditional form of the justificatory question is incoherent. Together with foundationalism, which gave rise to it, that question assumed an Archimedean platform outside the tribe and outside its history, a platform upon which an individual engaged in rational evaluation might stand. But for the god's-eye view such a platform would provide, the historically oriented approach has no need. Nor has it space in which such a platform might be placed. Though critical evaluation plays a vital role in the tribe's further development, criticism can only come from within the tribe.

I have dealt so far with three results of replacing belief with change of belief as the object of justification. First, evaluative criteria become comparative, eliminating the need to set a threshold for rational acceptability. Second, considerations relevant to the contexts of discovery and justification increasingly overlap, for what requires justification is now only what has been discovered, the set of statements with respect to which belief has changed. Third, as a result of these two, the problem of holism has dissolved, for it no longer makes sense to ask for the justification of any beliefs except those in which change is proposed. These changes are large, but more is to come. As my references to tribes and tribal membership may indicate, the *structure* of justification is not all that shifts with the transition to a developmental approach. So, too, I shall now argue, does the nature of the authority which underpins both belief itself and the process of belief justification. What has disappeared in the transition is not simply foundationalism and holism but methodological solipsism as well.

As seen by the static tradition, a belief's authority derived from the justificatory procedures to which it responded successfully, and any rational individual was in a position to administer the required tests. Though no one doubted that in practice many beliefs were accepted on authority— that of parents and teachers, for example—they need not have been. Each belief could, in principle, have been evaluated prior to its acceptance, and evaluation was therefore the only aspect of belief acquisition with which the philosopher needed to be concerned. All else could be left to psychology and sociology, depositories to which the tradition had already consigned considerations relevant to discovery. Within the developmental approach, however, science has ceased to be, even in principle, a one-person game and has become instead a social practice. It is now the group rather than the individual that guards the rationality of belief, and much of the belief which it guards, because constitutive of the community's way of life, is simply not subject to justification. The perspectives of the psychologist, the sociologist, and, above all, the anthropologist become relevant to philosophy after all.

I am, of course, now tracing a circle, from practice to constitutive beliefs and back again. But the circle is by no means vicious. It encompasses only the core of the beliefs and practices of a given tribe at a given time, and they, once grasped, determine a great deal else.[4] Enter the circle, as the historian must, at some selected time; reconstruct the beliefs that then determine the practice and the practices that then determine belief; and, with that temporal starting point established, watch the way in which beliefs and practice together develop from it. The starting point, just because it is the starting point, must remain a merely contingent set of historical states of affairs. But each step forward in time from it can be seen to be the product of choices which, the tribe's circumstances permitting, were made simply because they promised solutions to the problems which the practice in its current state was intended to resolve. Like its predecessor, the developmental approach to philosophy of science posits rational criteria for theory choice. But the criteria are now those of a tribe; assimilating them is part of what makes an individual a member of that tribe; and the criteria are therefore non-vacuously applicable only to an explanation of the further development of tribal practice, not to an understanding of its entire current state.

As to the criteria themselves, they remain much what they were before— closeness of fit, scope, simplicity, and so on, somewhat broadened by the addition of a few time-dependent standards, like observed fruitfulness. But it is now the group of initiates, not simply the rational individual, which bears the ultimate authority in their application. For the developmental approach, evaluation itself is an extended process. When a new theory is first proposed, reasons for adopting it are few and equivocal. Typically, for example, its scope will be far narrower than that of its established predecessor,

4. I am speaking of what is usually called the *hermeneutic circle*. Charles Taylor, "Interpretation and the Sciences of Man," *Review of Metaphysics* 25, no. 1 (1971): 3–51 [reprinted in his *Philosophy and the Human Sciences: Philosophical Papers*, vol. 2 (Cambridge: Cambridge University Press, 1985), 15–57], provides a splendid introduction to this way of speaking.

but it will succeed brilliantly with a few established problems that its predecessor has to date been unable to resolve. Under such circumstances, individuals fully committed to the established standards for rational theory choice may nevertheless differ about which theory to choose, because they disagree about the weight to be given to different criteria. Theory choice becomes a matter of judgment, and the differences between the judgments of rational individuals become vital to the health of science.

Suppose that rationality did, as the tradition has supposed, constrain all individuals concerned with a choice between theories to make the same decision on the same evidence. How strong must the evidence be to justify the replacement of a long-established theory by a recently suggested alternate? If the requirement is set high, then no newly proposed theory would be allowed time to demonstrate its strengths; if it is set low, no established theory would be given the opportunity to defend itself against attacks. Solipsistic method would stifle scientific advance. A judgmental decision procedure permits the community to distribute the risks that any choice of life-form must involve.

At the start of this lecture, sketching the traditional static approach to philosophy of science, I divided its principal concerns into two sets. First was the establishment of an indubitable empirical foundation upon which a structure of generalizations about nature might be built or against which such a structure might be checked. Second was [the] establishment of a method, modeled upon mathematical proof, which would provide a chain of indubitable links, coercive for all rational human beings, between that structure and its foundation. (Where those links proved probabilistic rather than certain, it was the evaluation of their strength that was to be coercive.) In the transition to the developmental approach the foundation has become simply the body of belief[s] shared by the members of a scientific community at some given time. What is linked to it by method, logic, and reason is no longer some higher-level set of generalizations but only the beliefs of later members of the same community, beliefs which have evolved through research and evaluative criticism from those held at the chosen initial time. And, finally, the nature of the evaluative criticism has become judgmental rather than coercive, its locus transferred in the process from the rational individual to the collective membership of a group committed to the established standards of science. The transformation is apparently complete! But only apparently. One central element of the tradition is still missing, and with it I shall close.

The missing element is language or, more precisely, that part of language in which the components of the empirical foundations of sound knowledge were traditionally to be expressed. Whether its form was that of a sense-datum language or some other, it was to be independent of all forms of belief and capable of expressing the minimal components of all human experience, experience which could not be doubted by anyone exposed to it and possessed of normal sensory apparatus. The existence of some such neutral but omnicompetent descriptive language was an essential element of the static tradition, prerequisite to the fulfillment of its principal claims. What has happened to it in the ongoing transition to a developmental approach?

After three centuries of fruitless effort, no one quite continues to expect that anything remotely like a sense-datum language will be found. But very little attention has been paid by historically oriented philosophers of science to the consequences of accepting the everyday language of scientists as adequate to the evaluation of their knowledge claims. Most seem to feel that, though the descriptive language used by scientists is inevitably somewhat constrained by the theories with which it is used, it still comes close enough to a language of pure description to serve the functions of that ideal. I believe that the ideal of neutral description is itself at issue and that its abandonment may prove the most profound of all the consequences of the turn to history. Let me now try, far too briefly, to indicate how that could be the case.

Start with the obvious. When one acquires a vocabulary, a lexicon, one acquires a highly developed tool suitable, among other things, for description of the world. More particularly, if in part metaphorically, one acquires a taxonomy: the names of things, activities, and situations which will need to be described, as well as the names of features that will be useful in identifying and describing them. To permit identification, furthermore, the pinning of names to the things they name, the process of lexical acquisition must also associate the names of things with the names of the most salient of the features used to describe them. Until this learning process has proceeded a certain distance, description cannot even begin. By the time it can, however, one has learned far more than a language useful in description: one has also learned much about the world to which that language applies. That aspect of lexical acquisition holds, I believe, as much for students in university science courses as for those in kindergarten classes. Both are learning about the world and the lexicon together. Neither can put the resulting lexicon to use until that two-faced learning process has passed a certain level. There is no neutral, purely descriptive vocabulary from which to start.

I have briefly departed from history but return to it at once. What makes these remarks about language important is that history displays repeated episodes in which the price of new knowledge has been a change in descriptive language. Among the beliefs acquired with the lexicon are many that one may later find good reason to change. The development of science turns out to depend on alteration, not only of what one says about the world, but also of the lexicon which one uses to say it. These required lexical changes are at the heart of the phenomena I once labeled *manifestations of incommensurability*. Because the use of certain words has changed, some of the statements that recur in the texts of an older science cannot be translated in[to] the language of a subsequent science, at least not with the precision required to understand why they were made. That is the problem that gives rise to what I previously described as the first of the historian's two tasks: recapturing the integrity of an out-of-date scientific tradition.

Reading the scientific writings of an earlier age, the historian repeatedly encounters passages that make no sense. The difficulty is not that they contain statements which are clearly wrong, for that is to be expected. But these statements seem both so unreasonable and so unmotivated that it is difficult to imagine how a person whose writings were elsewhere models of reason and intelligence could have written them. Faced with such passages, the historian's task is to show how they can be understood, how sense can be made of them. Usually, an essential step toward that end is discovering and teaching to readers long-discarded ways of using some of the words these nonsense passages contain.

Extended examples would be required to illustrate and clarify this point and its immediate predecessors, but I must be content here with a single simple example, a tiny piece from one of a series of illustrations I am discussing elsewhere.[5] Among the requisites to an understanding of Aristotelian physics is the realization that for Aristotle the concept which English translators ordinarily cover with the term *motion* refers not only to change of position but to qualitative changes of all sorts; for example, the maturation of an acorn into an oak, the transition from sickness to health, or the transformation of ice from solid to liquid. For an Aristotelian these are all exemplars of the same natural category, the class of motions. Though differences between them were recognized (there were subcategories of motion), the principal features of all these examples were the ones they shared as motions, and the principal generalizations about motion were the ones that applied to them all. That unification was achieved by conceiving all motions as changes of state, as transitions from something to something, as possessed of two end points.[6] The features most salient to the identification or specification of a motion were thus its two end points, initial and final, together with the time required to pass between them.

Clearly, that way of attaching the term *motion* to nature had to be changed before the word could be used in Newtonian physics. For Newtonians the term *motion* refers to a state, not to a change of state. Its salient features are speed and direction, the properties which characterize it at an

---

5. Two other examples together with a much fuller version of this one can be found in my "What Are Scientific Revolutions?," in *The Probabilistic Revolution*, vol. 1, *Ideas in History*, ed. Lorenz Krüger, Lorraine J. Daston, and Michael Heidelberger (Cambridge, MA: MIT Press 1987), 7–22 [reprinted as chap. 1 in *The Road Since Structure: Philosophical Essays, 1970–1993, with an Autobiographical Interview*, ed. James Conant and John Haugeland (Chicago: University of Chicago Press, 2000)].

6. For these concepts see books III and V of Aristotle's *Physics* [Aristotle, *Physics*, ed. and trans. P. H. Wicksteed and F. M. Cornford, 2 vols., Loeb Classical Library 228, 255 (Cambridge, MA: Harvard University Press, 1957)]. Two Greek terms are involved, *kinesis* and *metabole*, and modern English translations usually render them as "motion" and "change" [respectively]. There are no better words available, but only the second refers to more or less the same phenomena as the term it is used to translate, so that their juxtaposition hides a crucial difference between the concepts embedded in Aristotle's Greek and in modern English. As Aristotle uses the terms, every referent of *kinesis* is a referent also of *metabole*, but not vice versa. (The referents of *metabole* include coming-to-be and passing away; those of *kinesis* are restricted to cases of *metabole* in which something persists through change.) In modern English, on the other hand, motions may be the cause of changes but they need not be changes in and of themselves: a growing organism is ipso facto a changing organism, but a moving body need not be a changing one.

instant. The term no longer refers at all to such changes as the growth of an oak or the passage from sickness to health. These semantic alterations are just some of the many closely related respects in which the Newtonian vocabulary cuts up the world differently from the Aristotelian. And it is only within the new taxonomy which that vocabulary provided that the conception of inertial motion could arise. For an Aristotelian, the concept of an enduring linear motion, because it lacked end points, was a contradiction in terms.

Note that what is at issue here is not the right or wrong way to use the term *motion*. No linguistic convention can be right or wrong, nor can any conventional taxonomy. But, for a specified purpose, one convention can be more effective than another, a better means to a given end. One reason for the linguistic changes that underlie Newtonian physics is that the objectives of the study of motion had changed between Aristotle and Newton, and Newtonian language was a far more powerful tool with which to work toward the new goals. The new use of the term *motion* is only one of a number of interrelated changes that made that language so effective.

History of science offers countless examples of this sort, though few are so consequential as the transition from Aristotelian to Newtonian physics. Historians encounter them when, trying to understand an out-of-date text, they discover that success requires their using some set of familiar terms in unfamiliar ways, allowing them to supply a different taxonomy from the one characteristic of their modern equivalents. Having had that experience, the historically oriented philosopher may conclude with me that the scientific knowledge transmitted by the text was embodied not only in the statements about nature which it contains but also in the now out-of-date language in which those statements were cast. To permit a further set of discoveries, that language required reform, and the process has continued since. Evolution of language, including elementary descriptive language, is as much a part of science as the evolution of laws and theories. There is no such thing as pure or as mere description, and a fundamental feature of the traditional concept of scientific objectivity is therefore at risk.

Here I must stop, but not without a peroration. I have been sketching the first fruits of taking seriously the obvious fact that scientific knowledge is a product of human history. They are first fruits, subject both to further development and, in some cases at least, to possible decay. And they are, in addition, all of them, controversial. There is no guarantee that the movement which has produced them will survive. But I believe it will, and I predict that, if it does, more than philosophy of science will be changed. Since the seventeenth century, science has been the central example of sound knowledge. No significant transformation of our understanding of science can take place without transforming our understanding of knowledge as well. That transformation is, I believe, also underway.

The Presence of Past Science (The Shearman Memorial Lectures)

Abstract

Lecture I: Regaining the Past

Scientific knowledge can be properly understood only as the result of a historical process, which involves significant conceptual changes. Only if one understands why an older set of beliefs was held and what appeared to be evidence for it can one hope to understand the process by which it was given up and replaced. The historian of science approaches the past as alien, and aims to make sense of it as a quasi-ethnographer of past concepts and beliefs.

Three examples of such quasi-ethnographic history of science are discussed.

Section 1. If we understand Aristotle's physics as an integrated whole, with concepts different from ours, we will understand why Aristotle *had* to think that a vacuum is impossible.

Section 2. Volta's early diagrams of the electric battery seem erroneous when seen through the spectacles provided by a later physics, but they make perfect sense once we recover the meanings of key terms current at the time of Volta's writings.

Section 3. Planck's early work on the black-body problem should not be read from the standpoint of developed quantum theory; we need to understand that Planck's terms attach to nature differently than our terms.

Section 4. Previous examples show that a community's taxonomy supplies its ontology—it gives names for things that its world can and cannot contain. A taxonomy and a belief system form an inextricable mix. The history of science thus requires recovery of past beliefs as conceptually alien but plausible, once the past concepts are understood in their own historical context. The importance of this point is not solely historiographic, but philosophical as well, as the following lectures will show.

Lecture II: Portraying the Past

Incommensurability is best understood as characterizing not a barrier to communication among contemporary practitioners of science, but the experience of a historian struggling to understand a conceptually alien past.

Section 1. The historian's work requires learning the language in which the past knowledge was stated. This involves mastering the structured lexicon incommensurable with the historian's own.

Section 2. To acquire a lexicon is to learn things about the world. Users of the same lexicon must divide the world into the same natural kinds and identify the same objects and situations as falling under these kinds. Some natural-kind terms are common to all human languages, but some develop in response to needs and environments of particular communities, and thus vary across languages, cultures, and historical periods. Natural kinds are organized hierarchically, by similarity and difference relations; they do not allow overlaps. An anomalous object that equally resembles members of two distinct kinds threatens the accepted taxonomy; the likely solution is a lexical redesign—a conceptual change likely to eventually affect vast areas of the lexicon.

The need for historical quasi-ethnographic interpretation arises from a disparity between current and past taxonomies, which preclude truth-preserving translations. Terms like *true* and *false* function only in the evaluation of the choices made *within* a community that has an ontology of kinds and a lexicon in place.

Section 3. Specialized scientific lexicons can be mastered only by a restructuring of the learner's antecedent vocabulary, derived from common parlance. Learning involves both acquiring new meanings for familiar words and new beliefs about the world. Mastering scientific natural kinds often involves learning the laws that govern the behavior of their members; at the same time, understanding the content of the laws requires learning how scientific kinds are constituted and differentiated. Neither the laws nor the definitions of scientific natural kinds are analytic, and can change in time. Different individuals can acquire identically structured lexicons by different routes. The ontological question about what is there in the world is inseparable from the epistemic question of how referents of a term are to be picked out.

What one is committed to by a lexicon is a set of possible worlds, allowed by the ontology of the natural kinds that they share. Discovering the actual world among the members of the set of possible worlds is what scientists do during the periods of normal science. However, scientific development sometimes has to restructure some part of the lexicon and thus gain access to worlds that were inaccessible before.

Lecture III: Embodying the Past

What are the philosophical consequences and problems of adopting a historical perspective that sees past science as alien, and past lexical structures as incommensurable with one another and with our own?

Section 1. The problem of *bridgeheads*: How much commonality is required to explain the success of the historian in reconstructing the belief system of another time? It is argued that minimal bridgeheads are sufficient for understanding, and that understanding requires not translation but bilingualism.

Section 2. The problem of relativism: Can the truth or falsity of a belief about the world depend on the lexicon of the community within which that belief is held? If the same statement can be made in different lexicons, then it must have the same truth value in all of them. But this is not always the case. What is relative is not truth, but effability. Lexicons themselves cannot properly be labeled *true* or *false*, but they can be evaluated in other ways. Pragmatists were right on this point: lexicons are instruments to be judged by their comparative effectiveness in promoting the ends for which they are put to use, and the choice among them is interest-relative.

Section 3. The problem of realism vs. constructivism: Can talk of world change be heard as anything but the wildest of metaphors? Formulations in *Structure* may have been misleading. It is not the case that the community stood still while the world changed about it, so that a statement that was true before the change became false after it. In fact, both the world and the community changed together with the change of the lexicon through which they interacted. Categories of the mind are required to constitute experience of the world. This Kantian position differs from Kant's own in that the categories in question are not necessary and universal, but on the contrary, contingent, local, historically situated, and subject to change.

Section 4. Given the role of a lexical structure in constituting a world, how can a lexicon change? Aspects of a community's knowledge of the world are built into the structure of the lexicon, and novel experiences sometimes strain that built-in knowledge in ways that can be relieved only by lexical change. This claim is illustrated and argued for through brief historical vignettes (Aristotle, Galileo, Einstein).

Section 5. The problem of connections between present and past: If the lexicon of a past community makes its world foreign, how can that be *our* past? We need two types of history: hermeneutic, ethnographic history, which reveals incommensurability and has great philosophical value; and presentist or Whig history, which is required for formation of present identity, especially that of the scientists. Hermeneutic narratives allow us to understand the past, while Whig narratives allow us to use the past in the present. The two types of historical narratives are mutually incompatible, but each is necessary in its own context and for its own purposes.

THE PRESENCE OF PAST SCIENCE

Thomas S. Kuhn

The Shearman Memorial Lectures
University College, London
November 23, 24, 25, 1987

CONTENTS

Regaining the Past

In these lectures I return to a set of themes that a few contemporaries and I first raised for discussion some twenty-five years ago.[1] Our topic was the nature and authority of scientific knowledge, and we approached it with a shared conviction that long-dominant views on our subject might be drastically altered by closer attention to what scientists actually do. Data about the behavior of scientists, we found wherever we could: some drawn from our own experience; some from an embryonic sociology of science. But the main source of the data with which we bombarded traditional empiricist approaches to philosophy of science turned out to be historical examples of scientific advance. Though other sources might, we thought, have done as well, relevant historical studies were already at hand, and we felt competent to develop still others for ourselves.[2]

In retrospect, I think we were misled by seeing history as a source primarily of data. Case studies, especially those we prepared for ourselves, provided not only data but a perspective from which to view them. That perspective informed our data, but of its role in our work we were at best

---

1. I think particularly of Paul Feyerabend, N. R. Hanson, and Stephen Toulmin. For the problems to be considered below, the views of the first two were especially important.

2. The early chapters of Herbert Butterfield's *Origins of Modern Science* [London: G. Bell, 1949] were important to many of us, both for what they said and for the other studies to which they led. Particularly important among the latter were Alexandre Koyré's *Études galiléennes* [Paris: Hermann, 1939].

very dimly aware. Our situation was typical for would-be innovators. We worked too much on problems and with concepts evolved within the viewpoint we sought to supplant. Often we missed clues that might have pointed us toward alternatives to the tradition we criticized. Only relatively recently has it become possible to look with a clearer eye at the territory our work disclosed.

For much of the last decade I have been reexploring that territory, increasingly guided by the conviction that scientific knowledge can be properly understood only as a product of history, of a temporally and spatially continuous developmental process. These lectures focus on one product of that exploration: a set of problems concerning the nature and consequences of conceptual change. Though much discussed in recent years, these problems look different when viewed as consequences of the nature of history rather than of the facts history provides. To highlight that difference, I shall introduce my topic proper with a few brief and dogmatic remarks about the historical perspective itself. Evidence for them will emerge here and there as these lectures proceed.[3]

Seen from the viewpoint of the historian, all knowledge of nature emerges from prior knowledge, usually by extending, but sometimes by partially replacing, it. That generalization is as relevant to the so-called context of justification as it is to the context of discovery. To discovery, the prior body of knowledge supplies the conceptual tools, the manipulative techniques, and much of the empirical data required for the emergence of cognitive novelty. To justification, the same prior body of knowledge provides the only standard of comparison by which a candidate to succeed it can be judged. In the sciences, that is, the foundation for future knowledge is present knowledge, and there is no other foundation—more neutral, less contingently situated—to be had. Contributing to knowledge and evaluating contributions made by others are historically and culturally situated activities: no individual can engage in either until he or she has mastered

---

3. See also my "Scientific Knowledge as Historical Product," to appear in *Synthèse*. [Kuhn is referring here to *Revue de Synthèse*, a French journal in history and philosophy of science. The article was never published there, however. It was delivered as a lecture in Tokyo, in May 1986, and it appeared in *Shisō* in Japanese translation in August 1986 (see introduction to this volume, p. xiv). It is published here in English for the first time.—Ed.]

both the language of the community to whom the contribution is offered and also a number of that community's currently accepted truths. As a descriptive statement about the way science actually develops, this assertion of historicity is trivial. But I take its import to be more than factual; it is somehow deeply implicated in the nature of knowledge itself.

If I am right that the cognitive foundation of the science of one time is the science of the immediately preceding time, then two distinct tasks are involved in providing examples for philosophers of science to analyze. The second task is widely recognized and apparently unproblematic: each example must display the route from an older body of knowledge claims to an expanded or revised successor; narratives which trace such routes are the historian's primary product. Before a narrative can begin, however, historians confront a prior task: they must regain for themselves and their audience the past from which their narrative sets out; they must, that is, reestablish both an older body of knowledge claims and also the nature of its appeal. During this stage of their work, historians are like ethnographers striving to understand and describe the apparently incongruous behavior of the members of an alien culture.

This ethnographic aspect of history is far less widely recognized than its narrative successor, and these lectures are throughout concerned with problems it presents. The balance of today's lecture introduces these problems by presenting three examples of the ethnographic task which must precede the beginning of a historian's narrative.[4] Both individually and collectively, these examples will display the past as alien, and tomorrow's lecture—where I return to issues I once described with such terms as *incommensurability* and *partial communication*—asks about what makes the

---

4. These examples were first worked out in roughly their present form for the opening lecture in a series delivered at the University of Notre Dame in the fall of 1981. Revised for presentation in an independent lecture, they have recently been published as "What Are Scientific Revolutions?," in *The Probabilistic Revolution*, vol. 1, *Ideas in History*, ed. Lorenz Krüger, Lorraine J. Daston, and Michael Heidelberger (Cambridge, MA: MIT Press, 1987), 7–22 [reprinted as chap. 1 in *The Road Since Structure: Philosophical Essays, 1970–1993, with an Autobiographical Interview*, ed. James Conant and John Haugeland (Chicago: University of Chicago Press, 2000)]. I now think that [that] title [is] decisively misleading, and discovering the difficulties it raises has been an important learning experience for me, a topic to which I return briefly in lecture II.

past foreign and about the extent to which and the manner in which its foreignness can be transcended. Finally, my third lecture will confront some consequences of the position developed in the previous two, arguing that the threats often ascribed to that position—notably relativism and idealism—are either not pertinent or else not appropriate causes for alarm.

1

My first example is an experience—the start of my understanding of Aristotelian physics—which forty years ago first persuaded me that history of science might be relevant to philosophy of science. I first read some of Aristotle's physical writings in the summer of 1947, while a graduate student of physics trying to prepare a case study on the development of mechanics for a course in science for nonscientists. Not surprisingly, I approached Aristotle's texts with Newtonian mechanics clearly in mind. The question I hoped to answer was how much mechanics Aristotle had known, how much he had left for people like Galileo and Newton to discover. Given that formulation, I rapidly discovered that Aristotle had known almost no mechanics at all. Everything was left for his successors, mostly those of the sixteenth and seventeenth centuries. That conclusion was standard, even among those who knew Greek, which I did not, and it might in principle have been right. But I found it bothersome, because as I was reading him, Aristotle appeared not only ignorant of mechanics, but a dreadfully bad physical scientist as well. About motion, in particular, his writings seemed to me full of egregious errors, both of logic and of observation.

These conclusions were, I felt, unlikely. Aristotle, after all, had been the much-admired codifier of ancient logic. For almost two millennia after his death, his work played the same role in logic that Euclid's played in geometry. In addition, Aristotle had often proved an extraordinarily acute naturalistic observer. In biology, especially, his descriptive writings provided models that were central in the sixteenth and seventeenth centuries to the emergence of the modern biological tradition. How could his characteristic talents have deserted him so systematically when he turned to the study of motion and mechanics? Equally, if his talents had so deserted him, why had his writing in physics been taken so seriously for so many centuries after his death? Those questions troubled me. I could easily believe that Aristotle had stumbled, but not that, on entering physics, he had totally collapsed. Might not the fault be mine rather than Aristotle's, I asked myself? Perhaps not all his words had meant to him and his contemporaries quite what they meant to me and mine.

Feeling that way, I continued to puzzle over the text, and my suspicions ultimately proved well-founded. I was sitting at my desk with the text of Aristotle's *Physics* open in front of me and with a four-colored pencil in my hand. Looking up, I gazed abstractedly out of the window of my room—the visual image is one I can still recall. Suddenly the fragments in my head sorted themselves out in a new way, and fell into place together. My jaw dropped, for all at once Aristotle seemed a very good physicist indeed, but of a sort I'd never dreamed possible. Now I could see why he had said what he'd said, and why he had been believed. Statements that I had previously taken for egregious mistakes now seemed to me, at worst, near misses within a powerful and generally successful tradition.

That sort of experience—an increasing puzzlement and malaise suddenly resolved by a redescription, [a] resorting, and a reassembling of parts—often characterizes an early stage in the recovery of the past. Always it leaves much piecemeal mopping up to do, but the central change cannot be experienced piecemeal, one step at a time. Instead, it involves some relatively sudden and unstructured transformation in which some aspects of the ideas and behaviors under study sort themselves out differently and display patterns different from those visible before.

To make all this more concrete, let me now illustrate some of what was involved in my discovery of a way of reading Aristotelian physics, one that made the texts make sense. A first illustration will be familiar to many. When the term or terms rendered "motion" by translators occurs in Aristotelian texts, it refers to change in general, not just to the change of position of a physical body.[5] Change of position, the exclusive subject of mechanics

---

5. In fact, there are two terms which translators render as motion or sometimes as change: *kinesis* and *metabole*. All examples of *kinesis* are examples of *metabole* as well, but not conversely. Examples of *metabole* include coming-to-be and passing away, and these are no *kineses*, because they lack one end point. Cf. Aristotle, *Physics*, book V, chaps. 1–2, esp. 225a1–225b9. I shall here use *motion* for *kinesis*, excluding change from being to nonbeing and its converse.

for Galileo and Newton, is one of a number of subcategories of motion for Aristotle. Others include growth (the transformation of an acorn to an oak), alterations of intensity (the heating of an iron bar), and a number of more general qualitative changes (the transition from sickness to health). Aristotle recognizes, of course, that the various subcategories are not alike in *all* respects; but the cluster of features relevant to the recognition and analysis of motion are, for him, the ones applicable to changes of all sorts. In some sense that is not merely metaphorical, all these varieties of change are seen as like each other, as constituting a single natural family. Aristotle is explicit about the features they must share: a cause of motion, a subject of motion, a time interval in which the motion takes place, and two end points of the motion, those in which it begins and ends.

A second aspect of Aristotle's physics—harder to recognize and even more important—is the central role of qualities or properties in its conceptual structure. By that I do not mean simply that it aims to explain quality and change of quality, for other sorts of physics have done that. But Aristotelian physics inverts the ontological hierarchy of matter and quality that has been standard since the middle of the seventeenth century. In Newtonian physics, a body is constituted of particles of matter, and its qualities are consequences of the way those particles are arranged, move, and interact. In Aristotle's physics, on the other hand, the role of matter is secondary. Matter is needed, but only as a neutral substrate in which qualities inhere and which remains the same as those qualities change with time. That substrate must be present in all individual bodies, all substances, but their individuality is accounted for not in terms of characteristics of their matter but in terms of the particular qualities—heat, wetness, color, and so [on]—with which it is impregnated. Change occurs by changing qualities, not matter, by removing some qualities from some given matter and replacing them with others. There even appear to be conservation laws which some qualities must obey.[6]

---

6. Cf. Aristotle, *Physics*, book I, and esp. *On Generation and Corruption*, book II, chaps. 1–4 [Aristotle, *Generation of Animals*, trans. A. L. Peck, Loeb Classical Library 366 (Cambridge, MA: Harvard University Press, 1942)].

The notions of motion as change and of a qualitative physics are two possible points of entry to Aristotle's text. Either might have been discovered without the other: they are apparently independent. Nevertheless, as one recognizes these and other aspects of Aristotle's viewpoint, they begin to fit together, to lend each other mutual support, and thus to make a sort of sense collectively that they individually lack. In my original experience of breaking into Aristotle's text, the new pieces I am describing and the sense of their coherent fit actually emerged together. That recognition of coherence is a second characteristic aspect of the experience of regaining or recapturing the past. Indeed, *coherence* is too weak a word. As one's initial points of entry begin to fit together, other points seem to follow almost necessarily. Sometimes one predicts what an author must have believed, and then finds it stated later in the text.

A third aspect of Aristotle's physics will begin to fill in the relationships between those already introduced. In the absence of external interference most changes of quality are asymmetric, especially in the organic realm, which provides Aristotle's model for natural phenomena. An acorn naturally develops into an oak, not vice versa. A sick man often grows healthy by himself, but an external cause is needed, or believed to be needed, to make him sick. One set of properties, one end point of change, represents a body's natural state, the one it strives to achieve for itself and thereafter to maintain.

Taken together, these properties (more accurately, one of their proper subsets) constitute what has come to be called the body's essence.[7] Whether realized or potential, these essential properties make the body what it is. In particular, they provide the pattern for the body's natural development, the goal it strives by its nature to achieve: the maturation of the oak realizes the essence already present in the acorn. Changes of position

---

7. The term *essence* derives from medieval Latin translations of Aristotle: there is no entirely equivalent term in his Greek. But a concept like that of essence plays a central role in his physics, and modern translators often have good reason for introducing the corresponding term, usually as a substitute for *eidos* (more often translated [as] "form") or for *physis* (more often translated [as] "nature"). The absence of a term corresponding in full to the concept is an aspect of a more-than-verbal difficulty in Aristotle's position. More about it will be found in the next note.

display essence, too. The quality which a stone or other heavy body strives to realize is position at the center of the universe; the natural position of fire is at the periphery. That is why stones fall toward the center until blocked by an obstacle and why fire flies upward to the heavens. They are realizing their nature just as the acorn does through its growth. Given this notion of essence, the previously independent concepts of motion as change and of a physics of qualities become closely interrelated aspects of a single integrated viewpoint.

What underlies that interrelationship is the classification of a body's position or place as one of its qualities. Place at the center is to a stone what leaf size and shape [are] to the mature oak or what normal pulse rate is to the healthy man or woman. None of these qualities need be realized (the stone may be on a hilltop; an acorn has no leaves; pulse rate may be disturbed by illness). But all these bodies must be characterized by some quality of the relevant sort and must strive to realize the one that is natural to it. Making place one of the qualities is consequential.[8] The qualities of

---

8. The assertion that Aristotle considers place a quality is too categoric. His position is both elaborate and on occasions apparently inconsistent. The totality of qualities which must inhere in matter (*hyle*) to constitute a substance are the form (*eidos*) of the corresponding body. The question [of] whether or not place (*topos*) is a quality is the question [of] whether or not a body's *topos* is a part of its *eidos*. Aristotle gives two different answers corresponding to two different uses of *eidos*.

In the first of these the *eidos* of a body is all of its qualities at some particular time. Some, like the hair color of a man or animal, are accidental (*symbebekos*): they could be different, for example at different times, the particular substance remaining the same. Others are essential (*kath 'auta* or *to ti esti*), like heaviness in a stone or rationality in man: if they were different the substance would not be the kind it is. When change occurs with respect to *eidos*, it is this first sense of *eidos* that is involved, and it explicitly excludes place from *eidos*. *Kinesis* can occur with respect either to *poion* (quality), or to *poson* (magnitude, size), or to *topos* (place), and only the first of these is change with respect to *eidos*. In particular, Aristotle states that *topos* cannot be *eidos* because a body is not separable from its qualities but can separate from (move out of) its place. (Cf. *On the Heavens*, IV, 2, 310a24 [Aristotle, *De caelo*, trans. J. L. Stocks (Oxford: Clarendon Press, 1922)], and *Physics*, IV, 2, 209b23.)

But Aristotle also frequently restricts *eidos* to the essential or defining properties of a substance, those which make it what it is and which cannot change. (Cf. "to eidos to kata ton logon," *Physics*, II, 1, 193a30–33.) Conceived in this way, *eidos* is the formal cause of change, and the corresponding term is often used interchangeably with *physis* (nature) as a body's internal principle of motion. This is the use of *eidos* which supplies the concept later to be labeled *essence*, and in this use the *topos* which supplies the goal of a motion is part of the *eidos*, as potential position at the center is for a stone. When *topos* is used in this way Aristotle sometimes

a falling stone change as it moves: the relation between its initial and final states is like that between the acorn (or sapling) and the oak or between the youth and the adult. For Aristotle, therefore, local motion is a change of state, rather than a state as it is for Newton. Newton's first law of motion, the principle of inertia, then becomes unthinkable, for it is only states that can endure in the absence of external intervention. If motion is not a state, then an enduring motion requires force throughout.

One could continue for some time in this manner, fitting individual bits of Aristotelian physics into place in the whole. But I shall instead conclude this first example with a last illustration, Aristotle's view concerning the vacuum or void. It proves a particularly striking exhibit of the way in which a number of theses, apparently arbitrary in isolation, can together form a structure within which each and all find support. Aristotle says that a void is impossible: his underlying position is that the notion itself is incoherent. By now it should be apparent how that might be so. If position is a quality, and if qualities cannot exist separate from matter, then there must be matter wherever there's position, wherever body might be. But that is to say that where there is no matter there is no place: the concept of empty space becomes very [much] like a contradiction in terms, a close relative of the concept square circle. In Aristotle's words: "Since the void (if there is

---

speaks of it as the body's *auto topos* (own place) or of the *oikeios topos* (immediate place, the place at which the body happens to be). (Cf. *Physics*, IV, 4, 211a6, 211a29; V, 6, 230b27.) But this distinction between uses of *topos*, like the related one between uses of *eidos*, is not drawn regularly, and failure to draw it sometimes seems fundamental to Aristotle's argument.

Aristotle is using the second sense of *eidos* when he says, for example, "the movement of each body to its own place (*topos*) is movement to its own form (*eidos*)" (*On the Heavens*, IV, 2, 310a35). An even clearer example is: "For in general that which is moved changes from something into something, the starting-point and the goal being different in form (*eidei*). For instance, to recover health is to change from disease to health, to increase is to change from smallness to greatness. Locomotion must be similar: for it also has its goal and starting-point—and therefore the starting-point and the goal of the natural motion differs in form (*eidei*)" (*On the Heavens*, I, 8, 277a13–21, quoted from the Oxford translation by J. L. Stocks).

The difficulties of this double usage come to a focus in Aristotle's definition of place (*topos*) as "the boundary of the containing body at which it is in contact with the contained body" (*Physics*, IV, 4, 212a5). Place is here associated firmly with the body whose place it is and, simultaneously, made external to that body which can, Aristotle at once insists, move out of it.

any) must be conceived as a place in which there might be body but is not, it is clear that, so conceived, the void cannot exist at all."[9]

There are, of course, other ways to conceive the void, ways which eliminate its air of paradox, but Aristotle cannot choose among them freely. The one he does choose is largely determined by his conception of motion as change of state, and other aspects of his physics depend on it as well. If there could be a void, then the Aristotelian universe or cosmos could not be finite. It is just because matter and space are coextensive that space can end where matter ends, at the sphere of the stars, beyond which there is nothing at all, neither space nor matter. But if the universe were unbounded, then the rotating sphere that carries the stars would have to be infinite, a source of great difficulties for astronomy. More difficult still, in an infinite universe, any point in space would be as much the center as any other. There would, then, be no special position at which stones and other heavy bodies realized their natural quality. Or, to put the point in another way, in a void a body could not discover the location of its natural place. It is just by being connected with all positions in the universe through a chain of intervening matter that a body is able to find its way to the place where its natural qualities are fully realized. The presence of matter is what provides space with structure.[10] Thus, both Aristotle's theory of natural local motion and ancient geocentric astronomy would be jeopardized by the rejection of Aristotle's conception of the void. There is no way to "correct" it without reconstructing much of the rest of his physics. It is no accident that, in the event, the infinite Copernican universe, the mechanics of Galileo and Newton, and the first terrestrial vacua all emerged together.

2

Those remarks, though both simplified and incomplete, should sufficiently indicate how the pieces of Aristotle's description of the physical world lock

---

9. *Physics*, IV, 7, 214a16–20, quoted from the Loeb translation by Philip H. Wicksteed and Francis M. Cornford. As the end of the previous note indicates, an ingredient is missing from my sketch of this argument: Aristotle's definition of place, developed just before the discussion from which this quotation is taken.

10. For this and closely related arguments, see *Physics*, IV, 8, esp. 214b27–215a24.

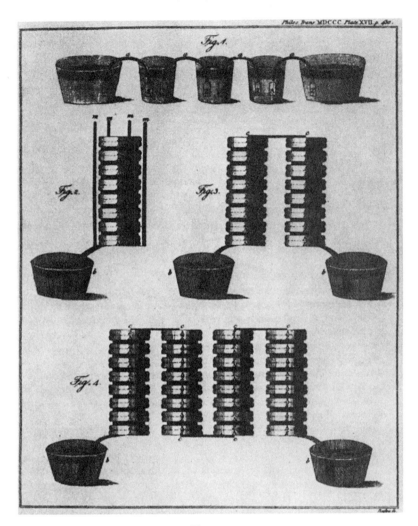

Figure 1

together to form an integral whole, one that had repeatedly to be broken and reformed in the historical development of the conceptual vocabulary that I, the initially ethnocentric historian, had tried to impose on Aristotle's text. Rather than extend them further, I shall proceed at once to a second example, this one situated at the beginning of the nineteenth century.

Among the notable events of the year 1800 is Volta's discovery of the electric battery, first announced in a letter to Sir Joseph Banks, president of the Royal Society.[11] It was intended for publication and was accompanied by the illustration reproduced here as figure 1. For a modern audience there is something odd about it, though the oddity is seldom noticed, even by historians. Looking at any one of the so-called piles (of coins) in the lower two-thirds of the diagram, one sees, reading upward from the bottom right, a piece of zinc, Z, then a piece of silver, A, then a piece of wet blotting paper, then a second piece of zinc, and so on. The cycle zinc, silver, wet blotting paper is repeated an integral number of times, eight in Volta's original il-lustration. Now suppose that, instead of having all this spelled out, you had been asked simply to look at the diagram, then to put it aside and repro-duce it from memory. Almost certainly, those of you who know even the most elementary physics would have drawn zinc (or silver), followed by wet blotting paper, followed by silver (or zinc). In a battery, as we all know, the liquid belongs *between* the two different metals.

Clearly, this problem of recognition results from looking at Volta's dia-gram through the conceptual spectacles provided by a later physics. But if one does recognize the anomalies in the diagram and puzzles over them with the aid of Volta's text, two related misreadings emerge for simultane-ous correction. For Volta and his followers, the term *battery* refers to the entire pile, not to a subunit composed of a liquid and two metals. Those subunits, furthermore, which Volta refers to as *couples*, do not literally in-clude the liquid at all. For him, the subunit is the two pieces of metal in contact. The source of its power is the metallic interface, the bimetallic junction that Volta had previously found to be the seat of an electrical ten-sion, of what we would call voltage. The role of the liquid is simply to con-nect one unit cell to the next without generating a contact potential which would neutralize the effect.

These features are all closely interrelated. Volta's term *battery* is bor-rowed from artillery, where it refers to a group of cannons fired together

11. Alessandro Volta, "On the Electricity Excited by the Mere Contact of Conducting Substances of Different Kinds," *Philosophical Transactions* 90 (1800): 403–31. On this subject see Theodore M. Brown, "The Electric Current in Early Nineteenth-Century French Phys-ics," *Historical Studies in the Physical Sciences* 1 (1969): 61–103. [Kuhn made the following note to himself: "[I] must add some remarks about this story being French, where mostly physi-cists worked on battery. In England things would be different. Acknowledgements to [June] Fullmer; cite Geoff Sutton."—Ed.]

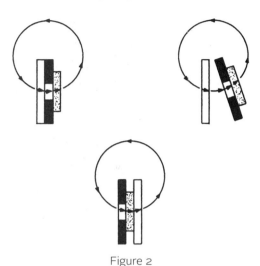

Figure 2

or in rapid succession. By his time it was standard to apply it also to a set of series-connected Leyden jars or condensers, an arrangement that multiplied the tension or the shock that could be gained from an individual jar acting alone. This electrostatic model is the one to which Volta is assimilating his new apparatus. Each bimetallic junction is a self-charging condenser or Leyden jar, and the battery is formed by their linked assembly. For confirmation, look at the top part of Volta's diagram, which illustrates an arrangement he called "the crown of cups." This time the resemblance to diagrams in elementary modern textbooks is striking, but there is again an oddity. Why do the cups at the two ends of the diagram contain only one piece of metal? What accounts for the apparent incompleteness of the two end cells? The answer is the same as before. For Volta, the cups are not cells but simply containers for the liquids that connect the bimetallic horseshoe strips or couples of which his battery is composed. The apparently unoccupied positions in the outermost cups are what we would think of as battery terminals, binding posts. The puzzling incompleteness was of our own making.

As in the previous example, the consequences of this view of the battery are widespread. For example, as shown in figure 2, the transition from Volta's viewpoint to the modern one reverses the direction of current flow. A modern cell diagram (at the bottom of the figure) can be derived from Volta's (upper left) by a process like turning the latter inside out (upper right). In that process, what was current flow internal to the cell becomes the external current, and vice versa. In the Voltaic diagram the external current flow is from black metal to white, so that the black is positive. In the modern diagram both the direction of flow and the polarity are reversed. Far more important conceptually is the change in the source of the current. For Volta, the metallic interface was the essential element of the cell and necessarily the source of the current it produced. When the cell was turned inside out, the liquid and its two interfaces with the metals provided its essentials, and the source of the current became the chemical effects at these interfaces. During the 1820s and 1830s, when both viewpoints were briefly in the field at once, the first was known as the contact theory of the battery, the second as the chemical theory.

Those are only the most obvious consequences of regarding the battery as an electrostatic device, and some of the others were even more immediately important. For example, Volta's viewpoint suppressed the conceptual role of the external circuit. What we would think of as an external circuit is simply a discharge path, like the short circuit to ground that discharges a Leyden jar. As a result, early battery diagrams in the Voltaic tradition do not show an external circuit unless some special effect, like electrolysis or heating a wire, is occurring there, and then very often the battery is not shown. Not until the 1840s do modern cell diagrams begin to appear regularly in books on electricity. When they do, either the external circuit or explicit points for its attachment appear with them. Examples are shown in figures 3 and 4.[12]

Finally, the electrostatic view of the battery leads to a concept of electrical resistance very different from the one now standard. There is an electrostatic concept of resistance, or there was in this period. For an insulating material of given cross-section, resistance was measured by the shortest length the material could have without breaking down—without leaking,

---

12. These illustrations are from Auguste [Arthur] de La Rive, *Traité d'électricité théorique et appliquée*, vol. 2 (Paris: J.-B. Baillière, 1856), 600, 656. Structurally similar but schematic diagrams appear in Faraday's experimental researches from the early 1830s [see Michael Faraday, "Experimental Researches in Electricity," *Philosophical Transactions of the Royal Society of London* 122 (January 1832): 130–31]. My choice of the 1840s as the period when such diagrams became standard results from a casual survey of electricity texts lying ready to hand. A more systematic study would have had to distinguish between British, French, and German responses to the chemical theory of the battery. [Kuhn indicated that he wished to change this footnote. Unfortunately, he did not specify how, or why.—Ed.]

Figure 3

Figure 4

that is, or ceasing to insulate—when subjected to a given voltage. For a conducting material, resistance was measured by the shortest length the material could have without melting when connected across a given voltage. It is possible to measure resistance conceived in this way, but the results do not conform to Ohm's law. If one is to make measurements that do, one must reconceive the battery and circuit on a more hydrostatic model. Resistance must become like the frictional resistance to the flow of water in pipes. Both the invention and the assimilation of Ohm's law required a noncumulative change of that sort, and that is part of what made his work so difficult for many people to understand and accept. His law has for some time provided a standard example of an important discovery that was initially rejected or ignored.

3

At this point, I end my second example and proceed at once to a third, this one both more modern and more technical than its predecessors. It involves a new interpretation, not yet everywhere accepted, of Max Planck's early work on the so-called black-body problem.[13] At the end of 1900 Planck applied to that problem a classical method developed some years before by the Austrian physicist Ludwig Boltzmann. Using Boltzmann's method, Planck was able to derive the now familiar black-body distribution law he had himself proposed a few months before. That derivation marks the historical origin of the quantum theory, a theory which breaks with classical physics by requiring that the energy of microscopic bodies be restricted to discrete levels between which it can change only by discontinuous jumps. Prepared by my earlier examples, none of you will be surprised to hear that Planck's derivation papers have for many years been read as containing those revolutionary concepts—discontinuity and the discrete energy spectrum—associated with his derivation by a later physics. But that ethnocentric reading, like the corresponding reading of Aristotle or Volta, presents special difficulties, and these are usually resolved by making Planck a sleepwalker who did not quite understand what he was doing.[14] For example, in those derivation papers Planck says nothing about discontinuous energy changes or a restriction on permissible energy levels, and his derivation, read literally, is incompatible with those concepts. Anomalies thus result from reading Planck through modern spectacles, and their presence suggests the need for a different way of reading, one that restores conceptual coherence to the texts.

To see how that can be achieved, look first, as with Volta, at the antecedent derivation on which Planck modeled his own. Boltzmann had been considering the behavior of a gas, conceived as a collection of many tiny molecules moving rapidly about within a container and colliding frequently, both with each other and with the container's walls. From previous works of others, Boltzmann knew the average velocity of the molecules (more precisely, the average of the square of their velocity). But many of the molecules

13. For a fuller account, together with supporting material, see my *Black-Body Theory and the Quantum Discontinuity, 1894–1912* (New York: Oxford University Press, 1978; repr., Chicago: University of Chicago Press, 1987). A briefer account of the main arguments will be found in my "Revisiting Planck," *Historical Studies in the Physical Sciences* 14, no. 2 (1984): 231–52, reprinted in the University of Chicago Press edition of the book.

14. Part 4 of the paper cited in the previous note includes two other examples of ethnocentric readings of the development of modern physics, develops further reasons for calling these readings ethnocentric, and suggests what is placed at risk by their rejection.

Figure 5

were, of course, moving much more slowly than the average, others much faster. Boltzmann wanted to know what proportion of them were moving at, say, 1/2 the average velocity, what proportion at 4/3 the average, and so on. Neither that question nor the answer he found to it was new. But Boltzmann reached the answer by a new route, from probability theory, and that route was fundamental for Planck, since whose work it has been standard.

Only one aspect of Boltzmann's method is of present concern. He considered the total kinetic energy of E of the molecules. Then, to permit the introduction of probability theory, he mentally subdivided that energy into little cells or elements of size ε, as in figure 5. Next, he imagined distributing the molecules at random among those cells, drawing numbered slips from an urn to specify the assignment of each molecule and then excluding all distributions with total energy different from E. For example, if the first molecule were assigned to the last cell (energy E), then the only acceptable distribution would be the one which assigned all other molecules to the first cell (energy 0). Clearly, that particular distribution is a most improbable one. It is far more likely that most molecules will have appreciable energy, and by probability theory one can discover the most probable distribution of all. Boltzmann showed how to do so, and his result was the same as the one he and others had previously gotten by more doubtful means.

Boltzmann's probabilistic technique was invented in 1877, and twenty-three years later, at the end of 1900, Max Planck applied it to an apparently rather different problem, black-body radiation. Roughly speaking, the problem is to explain the way in which the color of a heated body changes with temperature. Think, for example, of the radiation from an iron bar, which, as the temperature increases, first gives off heat (infrared radiation), then glows dull red, and then gradually becomes a brilliant white. To analyze that situation Planck imagined a container or cavity filled with radiation; that is, with light, heat, radio waves, and so on. In addition, he supposed that the cavity contained a lot of what he called *resonators* (think of them as tiny tuning forks, each sensitive to radiation at one frequency, not at others). These resonators absorb energy from the radiation, and Planck's question was: How does the energy picked up by each resonator depend on its frequency? What is the frequency distribution of the energy over the resonators?

Conceived in that way, Planck's problem was very close to Boltzmann's, and Planck applied Boltzmann's probabilistic techniques to it. Roughly speaking, he used probability theory to find the proportion of resonators which fell in each of the various cells, just as Boltzmann had found the proportion of molecules. His answer fit experimental results better than any other then or since known, but there turned out to be one unexpected difference between his problem and Boltzmann's. For Boltzmann, the cell size $\mathcal{E}$ could have many different values without changing the result. Though permissible values were bounded—they could not be too large or too small—an infinity of satisfactory values was available in between. Planck's problem proved different: other aspects of physics determined $\mathcal{E}$. It could have only a single value given by the famous formula $\mathcal{E} = h\nu$, where $\mathcal{E}$ is the resonator frequency and h is the universal constant subsequently known by Planck's name. Planck was, of course, puzzled about the reason for the restrictions on cell size, though he had a strong hunch about it, one he shortly attempted to develop. But, excepting that residual puzzle, he had solved his problem and his approach remained very close to Boltzmann's. In particular, the presently crucial point, in both solutions the division of the total energy E into cells of size $\mathcal{E}$ was a purely mental division made for statistical purposes. The molecules and resonators could lie anywhere along the line and were governed by all the standard laws of classical physics. The restriction on cell size did not, for Planck, imply a restriction on the energy of individual resonators; their energy changed continuously with the passage of time.

That way of reading Planck removes anomalies, and by doing so recovers a piece of the past. What is at stake in the reconstruction will be revealed by a sketch of what happened next. The work just described was done at the end of 1900. Six years later, in the middle of 1906, two other physicists argued that Planck's result could not be gained in Planck's way. There was a minor mistake in his argument. To make his derivation work, a small but absolutely crucial alteration was required. The resonators could not be permitted to lie anywhere on the continuous energy line but only at the divisions between cells. A resonator might, that is, have energy o, $\mathcal{E}$, $2\mathcal{E}$, $3\mathcal{E}$, and so on, but not $(1/3)$ $\mathcal{E}$, $(4/5)$ $\mathcal{E}$, etc. When a resonator changed energy, it did not do so continuously but by discontinuous jumps of size $\mathcal{E}$ or a multiple of $\mathcal{E}$.

After those alterations, Planck's argument was both radically different and very much the same. Mathematically it was virtually unchanged, a fact which has made it easy to read Planck's 1900 paper as presenting the revised argument, the one that still obtains. But physically, the entities to which the derivation refers are very different. In particular, the element $\mathcal{E}$ has gone from a mental division of the total energy to a separable physical energy atom, of which each resonator may have o, 1, 2, 3, or some other number. Figure 6 tries to capture that change in a way that suggests its resemblance to the inside-out battery of my last example. Here, too, the change is elusive, difficult to see. But also, once again, it is consequential. The resonator has been transformed from a familiar sort of entity governed by standard classical laws to a strange creature whose very existence is incompatible with traditional ways of doing physics. As most of you know, changes of the same sort continued for another twenty years as similar nonclassical phenomena were found in other parts of the field.

Those later changes I shall not attempt to follow, but instead conclude this example, my last, by pointing to one other sort of change that occurred at an early stage. In discussing the previous examples, I pointed out that the elimination of anomalies in the reading of a text requires changes in the way terms like *motion* or *cell* attach to nature. In this example the changes the historian must notice show in the words themselves. When Planck around 1909 was at last persuaded that discontinuity had come to stay, he switched to a vocabulary that has been standard since, one that highlights this altered view of the physical situation with which his theory dealt. Previously he had ordinarily referred to the cell size $\mathcal{E}$ as the energy element. Now, in 1909, he began regularly to speak instead of the energy quantum, for *quantum*, as used in German physics, referred to a separable and indivisible part, an atom-like entity that could exist in isolation. While $\mathcal{E}$ had been merely the size of a mental subdivision, it had not been a quantum but an element.

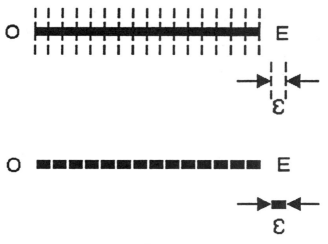

Figure 6

Also in 1909, Planck abandoned the acoustic analogy. The entities he had initially labeled *resonators* were now renamed *oscillators*, the latter a neutral term that refers to an entity that simply vibrates regularly back and forth. By contrast, *resonator* refers in the first instance to an acoustic entity or, by extension, to something that, like a tuning fork, responds gradually to stimulation, the amplitude of its oscillations increasing and decreasing continuously with the size of the stimulus applied. For one who believed that energy changes discontinuously, *resonator* was not an appropriate term, and Planck gave it up in and after 1909. With that change and the accompanying switch from *element* to *quantum,* essential aspects of Planck's black-body theory were embodied in the vocabulary since used to describe it. And it is through that vocabulary—used to translate Planck's talk of resonators and energy elements—that both scientists and historians have for more than half a century read his early papers, finding in them concepts which had not been invented when those papers were written.

4

That remark ends my third example. Rather than introduce still others, I shall quickly bring this lecture to a close by reminding you what they are examples of and pointing out, in however preliminary a way, a central feature which they all share.

At the start of the lecture I suggested that historians who study the development of knowledge of nature have a double task. They must provide a faithful explanatory narrative concerning changes in ideas about an interrelated set of natural phenomena. But before they can do so, they must undertake another task, less widely recognized. Narratives about development must open at some point or other in time, and historians must set the stage before telling the story of change from that point. They must, that is, show their audience what people then believed, and they may not do so simply by quoting what those people said. Literal quotation, as these examples show, often produces nonsense. Historians must instead approach older ideas in an ethnographic mode, one that aims to account for the coherence and plausibility of those ideas to the people who held them. Only if one understands why an older set of beliefs was held, what appeared to be evidence for it, can one hope to recount, analyze, or evaluate the process by which it was given up and replaced. The three examples I have discussed are concrete illustrations of both the need for and results of that quasi-ethnographic task. To this point their importance may appear to be solely historiographic, but in the remainder of these lectures, I shall try to show that it is philosophical as well. Let me briefly anticipate some of what I take that import to be.

In each of these examples, I have described a set of past beliefs about some aspect of nature. To do that, however, I have needed also in each case to describe the meanings of a few of the terms in which those beliefs were stated. These terms, furthermore, have been of a special sort generally; they are among the names of the taxonomic categories available to members of the speech community that uses them. They carry the community's ontology, supplying names for things which its world can and cannot contain. They are very like the terms that Mill described as the names of natural kinds, and his description of them, which I've only recently discovered, is increasingly influencing my thought.[15]

---

15. [John Stuart Mill, *A System of Logic, Ratiocinative and Inductive, Being a Connected View of the Principles of Evidence and the Methods of Scientific Investigation*, vols. 7–8 of *The*

In some cases, these problematic terms are still in use but with different meanings: the turn-of-the-century electrical terms *battery* and *resistance;* the Aristotelian terms for which translators substitute *motion* and *matter.* In other cases, the terms themselves were different but were used in ways that too easily suggest a misleading modern equivalent. *Oscillator* and *quantum* occur in contexts very like those which earlier contained Planck's *resonator* and *element,* but the later terms do not carry the same meaning as their predecessors; the standard substitution of the latter for the former misrepresents what Planck believed when his early black-body papers were written.

To promote clarity, my presentation of these examples has separated, so far as possible, descriptions of meanings from descriptions of belief. But the separation has been both incomplete and artificial. Throughout the interpretive efforts which produced these examples, beliefs and meanings were encountered together, in an inextricable mix. The puzzles with which each reinterpretation began took the form, roughly speaking, of statements of belief which could not be understood without rediagnoses of meanings. And for those rediagnoses these same puzzling statements of belief provided essential clues. That is why exercises of this sort are sometimes described as attempts to break into the hermeneutic circle.

In the next lecture I shall argue that such entanglements of beliefs and meanings are intrinsic to the nature of knowledge. Part of what counts as knowledge at any given time is acquired during the process of learning the language in which that knowledge is stated. Essential parts of a community's knowledge of nature are embodied in the structure of the lexicon which members of that community share. To acquire a lexicon containing such Aristotelian terms as *motion, place,* and *matter* is to learn things about the world: that a falling stone is like a growing oak, that nature can no more exhibit a vacuum or an inertial motion than it can a square circle. To describe Volta's battery with the lexicon of electrostatics (no other was then applicable to the phenomena involved) is to make the battery like the Leyden jar, to locate the source of current at the metallic interface, and to specify the direction of the current. To employ the terms *resonator* and

Collected Works of John Stuart Mill, ed. John M. Robson (London: Routledge and Kegan Paul, 1963–1974).—Ed.]

element as Planck did is to represent black-body radiation as like acoustic radiation and the energy element $\mathcal{E}$ as a subdivision in a continuum rather than as a separate energy atom. In each of these cases, the description of phenomena required commitment to a lexicon, and that lexicon brought with it restrictions on what those phenomena could and could not be. If nature were later found to violate those restrictions—as occurred in each of the examples I've presented—the lexicon itself would be threatened. Elimination of the threat required not simply the substitution of new beliefs for old but alteration in the lexicon with which the prior beliefs were stated.

Because lexical changes of that sort separate us from the past—in some areas even the quite recent past—we cannot fully recapture past science with our current lexicon. That is the position illustrated in my lecture today. Tomorrow, by sketching a model of how the lexicon works, I shall provide a more analytic base for this position, one that will suggest how the past is to be recaptured and what it is that I have been doing today. In the third lecture, my concern will shift to problems arising from positions taken in the preceding two, most notably problems of relativism, objectivity, and truth. If those problems are not yet apparent, let me leave you with a question that points toward them: Was Aristotle mistaken about the vacuum? When he said there could be no void in nature, was his statement simply false?

Portraying the Past

Early in my last lecture I suggested that before beginning a narrative about some aspect of the development of knowledge, historians must set the stage. Prerequisite to narrative, both for them and their audience, is a quasi-ethnographic interpretation of what was taken as knowledge in the community where and at the time when their narrative opens. Most of the lecture was then devoted to three examples of that initial interpretive enterprise, and the lecture's conclusion underscored a feature which all three examples shared. Each centrally involved a description, not only of the beliefs held by community members, but also of the meaning of some of the words in which those beliefs were expressed. Of those terms, some had since vanished from use. Others, though still in use, now functioned differently, carried different meanings. And until the older meanings were recovered, many passages in the texts which recorded an older knowledge seemed nonsense. Making them comprehensible required the study of meanings as well as of beliefs.

In taking up the alterations of meaning which initially separate the historian from the past, I am returning to an aspect of conceptual development which, a quarter century ago, Paul Feyerabend and I [independently] labeled incommensurability.[1] For me at that time, its primary application

---

1. I believe our resort to *incommensurability* was independent, and I have an uncertain memory of Feyerabend's finding the term in a draft manuscript of mine and telling me he had

was to the relation between successive scientific theories. In that application, changes in word meaning explained the characteristic difficulties of communication between proponents of competing theories. Corresponding conceptual changes were the basis for my talk of the gestalt switches which accompanied theory change. Looking back, that viewpoint seems to me correct in its essentials but in need of considerable modification of detail. One aspect of that modification is particularly relevant here: I have in the past modeled the experience of scientists moving forward in time too closely on the experience of the historian trying to move back.[2] It is the historian's experience, not that of scientists, that I illustrated in the examples developed last time, and it is in terms of the historian's experience that I want now to reintroduce the topic of incommensurability.

1

The term *incommensurability* is borrowed from ancient Greek mathematics, where it specified the relationship between two quantities which possessed no common measure—no unit, that is, which each contained some integral number of times. The hypotenuse and side of an equilateral right triangle are the most famous example; the radius and circumference of a circle are another. Applied metaphorically to the relation between successive scientific theories, *incommensurability* meant no common lexicon, no set of terms with which all components of both theories could be fully and

---

been using it too. But Feyerabend restricted incommensurability to language; I spoke also of difference in "methods, problem-field, and standards of solution," something I would no longer do except to the considerable extent that the latter differences are consequences of the lexical acquisition process, for which see below. The quotation is from *The Structure of Scientific Revolutions*, 2nd ed. (Chicago: University of Chicago Press, 1970), 103; it occurs also in the original (1962) edition.

2. Several difficulties result. First, the historian generally encompasses in a single leap a series of changes that took place historically in smaller steps. Second, it is a *group* (of scientists) that moves forward in time and an *individual* (the historian) who moves back, and the same descriptive terminology cannot uncritically be applied to both. An individual can, for example, experience a gestalt switch, but it is a category mistake to attribute an experience to a group. Mistakes of both these sorts have made it far harder than it need have been to describe the procedures available to the disputants at times of theory choice.

precisely stated.[3] Today, a quarter century after the appearance of Quine's *Word and Object*, *untranslatable* is a better word than *incommensurable* for what Feyerabend and I had in mind, and I shall be using it here.[4] Rather than claiming, for example, that the discussion of Aristotle's physics in my last lecture showed it to be here and there incommensurable with Newton's, I shall claim to have exhibited some Aristotelian beliefs that cannot be translated using the lexicon of Newtonian or of a still later physics.

I am, of course, just substituting one metaphor for another, but now the metaphor is Quine's rather than my own. It is truth-preserving translation that I am suggesting cannot always be done. Ordinarily it is impossible to substitute terms in current use for those in an older text in such a way that the truth value of each statement thus formed can properly be applied to its original.

To see what is involved, think briefly of translations of literature—of poetry, say, or drama. It is a cliché to say that these cannot be exact, that the words in the original language carry associations that only partially overlap those of their nearest equivalents in the language of the translation. Translators must therefore proceed by compromise, deciding in each case which aspects of the original it is most important to preserve, [and] which can, in the circumstances, be abandoned. About such matters, different translators may differ, and the same translator may, in different places, make different decisions about how to render a term, even though neither that term nor any of those that replace it is ambiguous. What I am suggesting is that the difficulties of translating science are far more like those of translating literature than has generally been supposed. In both science and literature, furthermore, the relevant difficulties arise not only when translating from one language to another but also when translating between earlier and later versions of the same language.

3. Here and elsewhere I speak of the lexicon, of terms, and of statements. But my concern is actually with conceptual or intensional categories more generally; e.g., with those that may reasonably be attributed to animals or to the perceptual system. Man's conceptualizing, taxonomizing ability is displayed most clearly in language, but there is more to language than conceptualizing, and conceptualizing is displayed also in ways that are prior to language.

4. That was not the case at the time we first publicly used *incommensurability*, in 1962, for *Word and Object* had appeared only two years before (Cambridge, MA: MIT Press, 1960).

The three examples in my last lecture exemplify these difficulties. In the absence of extended ethnographic interpretation—interpretation which transcends translations by summoning unfamiliar meanings of some of the terms they contain—each of my exemplary texts was systematically misleading. Occasional passages, clearly of central significance for their authors, patently failed to catch the meaning of the passages they replaced. Some sentences in those passages—sentences which must have been either true or false in the original—today read so strangely in translation that it is problematic whether what they appear to say can support truth values at all. It is sentences or statements of this sort that I have in mind when substituting *untranslatability* for *incommensurability*, statements for which no available translation techniques permit the preservation of truth value. When, from this point in these lectures, I speak of statements as translatable or untranslatable, it is truth-preserving translations I shall have in mind.

That point can be put in another way, though one that will later require much elaboration. It is widely assumed that anything which can be said in one language can be said in any other, at least if the lexicon of the language into which translation occurs is suitably enriched. This is the so-called linguistic effability thesis. If it were correct, then anything said in one language would carry its truth value with it when translated to another. Otherwise a statement could be true in one language, false when translated into another, a type of linguistic relativity that I shall insist is unacceptable. But another sort of linguistic relativity may not be. A statement that is a candidate for truth or falsity in one language may, in another, be impossible to state as a candidate for truth value at all.[5] I shall be arguing that something of that sort is the case. (That is why I asked, at the end of my last lecture, whether Aristotle had simply been mistaken when he proclaimed

---

5. Ian Hacking has introduced a similar distinction between types of relativism: the first would make the truth value of a proposition relative to a style of thought; the second would make relative only the availability of the proposition as a candidate for [truth or falsity]. Like me, he rejects the first as "inane subjectivism" but thinks the latter real. Hacking, however, sees the effects of changing thought styles as cumulative. See his "Language, Truth, and Reason," in *Rationality and Relativism*, ed. Martin Hollis and Steven Lukes (Cambridge: MIT Press, 1982), 48–66.

the impossibility of a void.) Though many of the statements that can be made with the lexicon of one language can be made also with the lexicon of another or with that of the same language at a later time, other statements cannot be carried over, even with the aid of an enriched lexicon.[6] The content of those statements can be communicated nevertheless, but what is required is not translation but language learning.[7]

That is what I had to engage in before I could understand the texts I discussed last time, and it is what I had to ask you to do here and there in following my discussion. During much of my last lecture, that is, I was speaking my own brand of everyday English, and with it I was able to communicate many of the beliefs of the scientists I discussed: Aristotle, Volta, and Planck. But not all their relevant beliefs could be communicated in that way. Where English versions of Aristotle's texts employed terms like *motion* and *place*, or *matter, form,* and *void*, I had first to supply some or all of you with unfamiliar meanings for those familiar terms, and then to use the restored versions to communicate what Aristotle believed. Similar alterations were required for passages in which Volta used terms like *battery* or *electrical resistance* and in which Planck used terms like *oscillator* and *energy element*. None of these terms applied to natural phenomena in the same way as their later replacements, and [the] lexicon in which those later terms occur cannot be used to provide words or phrases that can substitute for them.

Nor could our modern lexicon have been used to store those older terms except in the sense in which it can store special terms of art like

6. Compare my "Possible Worlds in History of Science," in *Possible Worlds in Humanities, Arts, and Sciences: Proceedings of Nobel Symposium 65* [ed. Allén Sture (Berlin: Walter de Gruyter, 1989), 9–32; reprinted as chap. 3 in *The Road Since Structure*]. I there argue that different lexicons give access to different sets of possible worlds.

7. It may be illuminating to apply this position to Quine's. His imagined anthropologist, the radical translator, is in fact not a translator at all but a language learner. Quine simply takes effability for granted: if the native language can be learned by the anthropologist, then it can, he supposes without comment, be translated into the language the anthropologist brought from home. Examining the sort of evidence available to the translator, he then argues for the indeterminacy of translation, but most of his arguments could equally well be read as pointing instead to its impossibility. Those arguments may well show that universality and determinacy of translation are incompatible, but they have no apparent bearing on the question of which one must be set aside.

Goodman's *grue* and *bleen*. As I yesterday emphasized in closing, the terms that I had to teach you were the names of fundamental taxonomic categories, bearers of the community's ontology. They functioned as projectible terms, the sorts of terms that can appear in laws of nature, counterfactual conditionals, or candidates for inductive generalization. They had, in other words, the characteristics of natural-kind terms, and two of those characteristics are presently essential. First, terms that refer to distinct natural kinds—whether to dogs and cats, gold and silver, or stars and planets— cannot overlap in their referents unless one of the kinds entirely contains the other. No object can, that is, be a member of two distinct kinds unless the two relate as genus to species. Second and equally fundamental, terms whose referents have, or are thought to have, such characteristics bear a special label *in the lexicon*, a label that indicates what may be expected of them.[8] It follows that, because of their overlap in reference, no pair of terms like the Aristotelian term *motion* and the Newtonian term *motion* can serve simultaneously as natural-kind terms. If both occur in the same lexicon, then at most one may bear the label that marks it projectible, makes it a term suitable for occurrence in natural laws. If included in the Newtonian lexicon, the Aristotelian term must be deprived of that label, and it then ceases to carry the meaning it had before.

2

In their present form these assertions are a promissory note that I cannot hope fully to redeem in these lectures. But I can hope to clarify them

---

8. I follow Mill in taking both the no-shared-members criterion and the natural-kind label as necessary features of natural-kind terms (those which denote Mill's *infima species*). Another, which Mill also particularly emphasizes, is that the members of natural kinds must share an indefinitely large number of features, some known at any particular time, others remaining to be discovered. When I speak of natural-kind terms, I have in mind any terms whose referents have these characteristics. I therefore include a term like *motion* (in Aristotle's sense, where the contrast is with the rest, but not in Newton's). [J. S. Mill, *A System of Logic, Ratiocinative and Inductive, Being a Connected View of the Principles of Evidence and the Methods of Scientific Investigation*, in *The Collected Works of John Stuart Mill*, ed. John M. Robson, vols. 7–8 (Toronto: University of Toronto Press; London: Routledge and Kegan Paul, 1963–91) book I, chap. vii, §§ 3–6; III, xxii, §§ 1–3.]

enough to permit discussion of some problems they present. Let me begin by developing and illustrating a preliminary model of the lexicon as it is embodied by individual members of a language community and also (not the same thing) as it is embodied in the language community as a whole. Though schematic and simplistic, the model illustrates characteristics which any more articulated version should display. In particular, it is responsive to the puzzles of textual interpretation sketched in the earlier parts of these lectures, puzzles from whose contemplation it, of course, derives. Also, it points the way toward a view of meaning which relates the sense of a term to the way in which its referent is determined, without in the process succumbing to the difficulties that have beset verification theories.

My concern is with the part of the lexicon that contains putatively referring terms (mostly nouns), each linked to the names (mostly adjectives) of properties or features that are of use in picking out its referents. Such a lexicon embodies the taxonomy of the language community whose members use it. It names the kinds of things, of behaviors, and of situations which occur in their natural and social world, and it also names the more salient characteristics of those kinds, the characteristics by which they are known. The knowledge it embodies is thus about both language and the world, about the names of things and properties, on the one hand, and about those same things and properties, on the other. Doubtless its evolutionary origin is prelinguistic (humans are not the only animals that deploy taxonomies), but I am here concerned with the form it takes when incorporated into language.

Some categories in the lexicon must be innate, genetically determined, shared by all human creatures. Individuated physical things are a likely example, as are focal aspects of perceived color.[a] Other categories, though not innate, are likely to be species universal by virtue of shared aspects of the natural environment. It is hard to imagine a language which has no word for the Sun, no words for night and day, no word whose referents include the stars. But still other lexical categories have evolved in response to the developing needs of particular communities, and these may vary with both place and time, with the relevant community's environment, and with the way members of the community interact with that environment.

Categories of this sort may differ from culture to culture, language community to language community, or, within a given community, from one historical period to the next.[9] Transmitted from generation to generation as part of the process of linguistic socialization, these ever-developing differences limit the extent of communication possible between members of groups with different lexicons. The same differences restrict communication with the past. In particular, they limit the specificity with which past scientific beliefs can be stated with the current lexicon.

Let me give an example of how I conceive the lexicon and lexical change. Compare the taxonomy used in Greek antiquity to categorize observations of the heavens with the revised version employed from the mid-seventeenth century. In antiquity there were just two sorts of celestial bodies, planets and stars. Most of them were visible as points of light. All of them were eternal, visible most nights of the year, and in regular motion during and between appearances. Other phenomena seen in the night sky—comets, shooting stars, the Milky Way—shared few of these features and were placed in the separate, noncelestial category of meteors, itself further subdivided. Of the celestial bodies themselves, planets were distinguished from stars by a number of further features. They tended to be brighter than stars, to appear only in the zodiacal region of the heavens, and to shine more steadily than the twinkling stars. More salient still, though stars and planets move steadily together in westward circles around the celestial pole, planets possessed an additional, much slower eastward motion through or among the stars. Utilizing these distinguishing features together, the Greeks identified seven planets: the Moon, Mercury, Venus, the Sun, Mars, Saturn, and Jupiter.

Like the examples I developed last time, this one deals with the taxonomy of a separable area of natural phenomena, in this case heavenly phenomena. Like them also, it is about both words and things in an inextricable mix. (That is why my talk about the lexicon so often takes the

---

9. Chomskian arguments for the universality of language seem to me entirely persuasive with respect to syntax and perhaps also compositional semantics. [See, for example, Noam Chomsky, *Language and Mind*, 2nd ed. (New York: Harcourt Brace Jovanovich, 1972); an updated, 3rd ed. was published in 2006 by Cambridge University Press.] But I see no evidence that they apply also to the lexical semantics of individual terms and phrases.

form of talk about the world and vice versa, an apparent bit of sharp practice which I mean to examine at the end of the hour.) In this case, too, the
words involved are of two sorts. On the one hand, they name kinds, mostly
natural kinds (planets) but also artefactual kinds (batteries) which satisfy
the same conditions. On the other, they name properties by which the referents of the first set of terms can be picked out. A novice learning how
the community categorizes a given area of natural phenomena may deploy
properties and terms for them previously available to him from other applications. In the case of the heavens, motion and circularity, day and year,
are likely examples. But other useful properties (twinkling, for example)
can be new. Learning to recognize these properties is part of acquiring the
taxonomic system they help to constitute.[10]

Note next, though my astronomical example is not altogether suited
for the purpose, the way in which these salient properties fulfill their taxonomic function. They supply useful guidance to classification, but they
don't need to specify necessary and sufficient conditions for class membership. Not all the bodies the Greeks took to be planets appear as points in
the sky; nor are they all especially bright; nor are they the only bodies that
wander slowly and, for a time, regularly among the stars. Rather than supply necessary and sufficient conditions, these properties supply a feature
space within which celestial bodies cluster like to like. Judged with respect
to these properties collectively, any planet was more like some other planets than it was like any star.[11] A previously unnoticed body in the night sky
could be classified by comparing it, *in the relevant feature space*, to bodies
whose classification was already known.

10. Not all the features useful in recognizing objects or kinds need have corresponding
names. Think of the features which permit facial recognition, for example, or the everyday ability to distinguish cats from dogs.

11. The resemblance to Wittgenstein's discussion of games is not coincidental. But Wittgenstein does not, I think, sufficiently stress the role of differentia and contrast sets. The ability to
pick out games depends not only [on] a knowledge of the features which games tend to share,
but also on [a knowledge of] those required to distinguish between some games and some
examples of, say, combat. [Ludwig Wittgenstein, *Philosophical Investigations*, trans. G. E. M. Anscombe (Oxford: Basil Blackwell, 1953); 4th ed., ed. P. M. S. Hacker and Joachim Schulte, trans.
G. E. M. Anscombe, P. M. S. Hacker, and Joachim Schulte (Oxford: Wiley-Blackwell, 2009).
Kuhn may have had in mind §§ 66–71, §75.]

At this point I am in danger of being taken for an advocate of a cluster theory of meaning or categorization: an object belongs in a given category if and only if it manifests *enough* of that category's defining features. But the position differs from mine in two respects. First, the features which I am invoking do not attach uniquely to individual categories. Rather they provide a space in which to determine the membership of a whole set of interrelated categories which, for that reason, have to be acquired together. They do supply information about features that the members of a given category tend to share, but a more significant role is often played by features with respect to which members of different categories differ. These differentiating features open empty space between the regions occupied by the members of the various categories. Second, and more important, because of that empty space, there is no need to specify the number of features a body must manifest to belong to a category. That is the significance of our dealing with terms that are marked in the lexicon as the names of natural kinds.

Natural kinds, I remind you, cannot overlap or even touch: no object can be a member of two different natural kinds unless the two relate as genus to species. Therefore, while nature continues to display the behavior which the lexicon was evolved to describe, any object actually encountered will, in the feature space supplied by the lexicon, clearly resemble members of one kind more than it resembles members of any other. An anomalous object, one that equally resembled members of two distinct kinds, would threaten the status of the natural kinds whose members it resembled. If the threat continued, the likely outcome would be lexical redesign.[12] Knowledge of nature is, as I have said before, embodied in the lexicon. When that knowledge is threatened, the lexicon itself, not just particular beliefs that can be stated with its aid, is in jeopardy.

---

12. Note that the concept of a natural kind is modeled on that of a biological species and traceable to Aristotle's discussion of secondary substance in *Categories* V. [See Aristotle, *Categories. On Interpretation. Prior Analytics*, trans. H. P. Cooke and Hugh Tredennick, Loeb Classical Library 325 (Cambridge, MA: Harvard University Press, 1938).] The threat posed by the overlap or contact of distinct natural kinds has a parallel in the threat posed to the concept of species by the theory of evolution. Two kinds that touch cease to be natural: a species that evolves into two species ceases to be a species.

Before further developing this model of the lexicon, it will help to examine the effects of lexical redesign. For the purpose, contrast Greek taxonomy for the heavens with the taxonomy that resulted from the work of Copernicus, Kepler, Galileo, and Newton. In the latter, the Sun has become a star, the Earth has joined the planets, and a new category, satellite, has been created for the Moon and the newly discovered satellites of Jupiter. At the same time, comets and the Milky Way (but not shooting stars) have been moved from the class of meteors to that of heavenly bodies. There has, in short, been a major taxonomic reshuffling, a change as to which bodies are like each other and which different.

What permits or accompanies that change is a set of changes, most of them small but with disproportionate effects, in the feature space within which the various heavenly bodies are grouped into clusters of different kinds. In the case of the stars, all the old features remain relevant, but a new one of special salience has been introduced: self-luminosity, without which no object is a star. In the case of planets also, the old features remain relevant, but the additional motion which makes planets wanderers among the stars has been reduced in salience, thus releasing the Moon to be classified as a satellite and the Sun to be classified as a star. Those reclassifications depend, however, on the addition of a second new feature, theoretical and hard to apply, like self-luminosity, but nonetheless crucial. Unless it circulates around a star, no object is a planet. To accommodate those wanderers that circulate instead around a planet, the class of satellites has been opened. Other changes, some associated with the introduction of the telescope, are also relevant to the emergence of a post-Copernican astronomical feature space, but these will suffice for present purposes.

My reference to self-luminosity and to star-centered orbits as "theoretical and hard to apply" introduces features that are not directly accessible to the senses, and I mean shortly to say something about how such features are acquired and put to use. But let me first describe one additional characteristic of this lexical model. To this point, I have suggested that what underlies the early modern regrouping of heavenly bodies is a change in the features used to categorize them. But, clearly, not just any change of features would have produced regrouping. If nature had been otherwise, the continued practice of Ptolemaic astronomy might have added to and refined the feature space of Greek astronomy without any reshuffling of heavenly bodies among categories. Galileo's telescope might not, for example, have revealed satellites about Jupiter or resolved the Milky Way into a myriad of distinguishable stars. If that had occurred, there would have been no occasion to speak of lexical change.

What characterizes a lexicon is, thus, not the features it deploys, but the groupings that result from the use of those features, whatever they may be. What changed during the transition from ancient Greek to early modern astronomy were first and foremost the similarity-difference relations between objects in the heavens. In antiquity the Sun and Moon were like Mars and Jupiter; after Copernicus and Galileo, the Sun was like the stars, and the Moon was like the newly discovered satellites of Jupiter. But the same groupings may be achieved in many different feature spaces. In principle, no two members of a language community need use any of the same features in order to group the objects in their environment into the same sets of natural kinds. In practice, many of the features that they use doubtless are the same, but they need not be.[13] That is a third respect, perhaps the most important of all, in which my position differs from that of cluster theorists.[b]

Different individuals cannot, however, use just any features at all. Two characteristics must be shared by members of a speech community if they are to divide the phenomenal world into the same natural kinds, identify the same objects and situations as members of those kinds, and employ these identifications in their interactions with the world and with each other. First, in their individual embodiments of the lexicon, the same terms must be tagged with the natural-kind label. And second, whatever features their individual lexicons embody, each of the feature spaces that results must yield the same hierarchical relations among kind terms and the same similarity-difference relations among the referents of terms at the same hierarchical level. (They must, one might say, share kinship relations.) On this view of the lexicon, the meaning of a term is associated, not with any particular set of features, but with what I shall henceforth refer to as the

---

13. What is at issue is not whether they can identify the same features but whether they use the same ones in picking out the referents of a given term.

lexicon's *structure*, the hierarchical and similarity-difference relations that it embodies. What separates the language of early modern astronomers as a group from that of their Greek predecessors is not so much that they used different features to pick out the referents of terms like *planet* and *star* but that those terms occurred in lexicons with relevantly different structures.

It is structural difference, I suggest, that impedes truth-persevering translation. Historians, for example, regularly report (I have often done it myself) that the Greeks said, "The Sun is a planet," and there are good reasons for them to do so. Our lexicon permits no closer rendering of the original. Forced to settle for the imperfect, translators of science, like translators of literature, do the best they can. But the translation is worse than imperfect. Using our lexicon, "The Sun is a planet" is false. The Greeks, we therefore suppose, were mistaken. But in the Greek lexicon, the Sun *was* a planet; it was, that is, more like Mars and Jupiter than like any of the stars. The corresponding Greek sentence was therefore true, not simply believed to be true.

My point, let me say again, is not that one and the same sentence was true for the Greeks and is false for us. Rather I am saying that, though the two word strings are the same, the statements made with them are different, and that there is no truth-preserving way in which the Greek statement can be rendered in our later lexicon. In particular, it will not do to substitute for the Greek equivalent of *planet* some string of terms providing a putative definition in terms of features used by the Greeks. There is no such string: different Greeks could have used different features. In any case, the features used by an individual Greek provided a system of interrelationships, not a meaning for single terms. And, finally, the Greek use of *planet* required that it occur in the lexicon as the name of a natural kind, and no feature string in a modern language can supply that marker without inconsistency. Adding it to the lexicon would, for example, make the Sun a member of two intersecting kinds.

3

I shall return to these issues of truth and translation, but want first to extend this model of the lexicon to terms and features of a more elaborate sort. The very considerations that led to the choice of this astronomical example limit the uses to which it can be put. Particularly where it dealt with antiquity, the example was restricted to a vocabulary of objects and properties that come as close as vocabulary ever does to a pure observation language. In fact, it was not pure: the division between stars, planets, and meteors can be made in other ways; the particular taxonomy implemented by the Greeks was not dictated by observation alone. But the Greek taxonomy could probably have been acquired by pure ostension, by a teacher's or a parent's pointing to exemplary objects in the night sky and simultaneously uttering the appropriate names. No words other than the names of celestial kinds need have been uttered. A large number of natural-kind terms must be learnable in this way, by direct exposure to the world in the presence of a guide who already knows both how the community categorizes the world and what it names its categories. Though not given by observation alone, those terms supply an important part of the community's observation vocabulary.

Like the astronomical example, the three examples in yesterday's lecture illustrate natural-kind taxonomies embodied in the lexicon as well as the way those taxonomies change with time. In addition, though I scarcely developed the point, those examples display the existence of feature spaces within which the referents of various natural-kind terms clustered. But few of the natural-kind terms in those examples could have been learned by direct ostension (think of Aristotelian matter and form, of the electric current, or of Planck's energy element, $\mathcal{E}$). To pick them out, to recognize even the features with respect to which they clustered, required the use of special instruments (e.g., the galvanometer) or of special calculation (Planck's energy element). In the acquisition of these terms and of the corresponding features, more than pure ostension was required. Pointing or its equivalent often played a role, but antecedently understood words were needed as well. And when that occurs, when words are required to introduce natural-kind terms, then the knowledge embodied in the lexicon includes more than knowledge of the kinds that do and do not exist and of the properties those kinds are likely or unlikely to share.

Imagine yourself in school learning the kind term *electricity*. Probably, though there is an alternative, you would require exposure to a variety of situations which manifest the presence of the referent of the term. Among them might be: a rubbed glass rod attracting chaff; the leaves of an electroscope diverging and reconverging as the rubbed rod was brought near to and then carried away from its knob; the mutual repulsion of two rubbed glass rods; the failure of the electroscope's leaves to reconverge if the rod made actual contact with the knob; a small spark passing from the rod to ground, and a reference to the larger spark known as lightning. In addition, you would require exposure to situations that might seem to manifest electricity but did not. A magnet attracts iron filings, but not chaff. And so on.

Exhibits like these are an especially effective way of teaching natural-kind terms, which is why they play so prominent a role in scientific education. But they are not absolutely required. Rather than being exhibited, the exemplary situations can be described to the student verbally using previously familiar terms, much as I have just done here. But, whether exposure is by exhibit or description, each exemplary situation must be accompanied by the utterance of one or more sentences containing the word *electricity* or some term like *electrified* transparently related to it. And some of these sentences must also include such other new natural-kind terms as *charge*, *conductor*, and *insulator*, terms which must be learned with *electricity* if any of them is to be learned at all. Appropriate sentences would include: "Rubbing a glass rod charges it with electricity." "An electrically charged body attracts uncharged bodies." "Two charged bodies attract each other." "Electricity travels through the conductor to which the leaves of the electroscope are attached." "Because glass is an insulator the glass rod does not lose its charge to ground." And so on.

Notice now that phrases like "attracts chaff," "discharges an electrified body to ground," and the like are the names of features in the electrical lexicon. What I have just exposed you to are some of the techniques by which one can acquire a feature space within which the referents of such natural-kind terms as *electricity*, *conductor*, and *insulator* can be identified. As in the case of terms like *star* and *planet*, no set of these features need supply necessary and sufficient conditions for the use of these electrical terms, and no two members of the community of electricians need to be exposed to the same examples or use the same features in picking out their referents. But all members of the community must use features that structure the

lexicon of natural-kind terms in the same way, that identify the same bodies as conductors or nonconductors and the same situations as manifesting
or failing to manifest the presence of electricity. Otherwise there is likely to
be trouble, for example with the installation of lightning rods.

Clearly, however, anyone who has acquired these terms in this way has
learned a great deal about electricity in the process. Having acquired the
requisite lexicon, one knows not only that electricity exists, but also that
electrically charged bodies repel each other but attract bodies that have
no charge; not only that conductors and insulators exist, but [also] that a
grounded conducting body touched to the knob of a charged electroscope
will cause its leaves to converge. Again, no two members of the community
need know all the same things about electricity. They may have acquired
their individual lexicons by different routes, using different examples.
But if their lexicons have the same structure, then each is equipped, without further changing that structure, to receive and assimilate the other's
knowledge.

In discussing the battery last time, I pointed out that a lexicon like the
one just described was in place before Volta's discovery, a lexicon that contained not only the terms *electricity, insulator,* and *conductor* but also *tension, condenser,* and a few others of the sort. That was the lexicon of the
community of electricians in 1800, and Volta fitted it to his discovery (or
his discovery to it) in a most straightforward way. What he had discovered,
he said, was a way to build and interconnect self-charging condensers. That
description was not simply wrong. It precisely fit the instrument Volta had
constructed. But that instrument is not quite the one we call a battery, and
the difference proved consequential. For example, the direction of current
flow in Volta's battery was opposite to its direction in ours. Interconnecting
his cells to magnify their effect required liquid conductors, while metallic
conductors are used between ours. That chemical effects occurred in these
connectors was for him incidental, a side effect ultimately to be eliminated
by the development of dry cells.

In the event, chemical effects could not be eliminated. Volta's assimilation of his discovery to the prior lexicon induced expectations that could
not be realized, and their elimination required more than correction of
an individual belief. His instrument itself had to be reconceived and that

change brought with it alterations of the feature space within which batteries could be singled out. (Volta's batteries, for example, lacked terminals
at which to attach an external circuit, perhaps the most salient of all the
features by which laymen today identify batteries.) During the 1830s and
1840s the list of features relevant to the electrical lexicon expanded, and the
structure of the kind terms in that lexicon changed. Most terms from the
vocabulary for static electricity—*charge, tension, conductor, condenser,* and
insulator—retained their old interrelationships. But *battery* was relocated
at a great distance from *condenser;* the old terms *resistance* and *conductibility*
were moved as well; and the structure was further adjusted to make room
for such new terms as *circuit* and *terminals.*

After those structural changes in the lexicon, much care was required
in reading older electrical texts. Volta's belief that chemical effects would be
eliminated from his liquid connectors could then be seen to be false, and it
had been false even at the time that Volta held it. But Volta's statement that
current flowed from zinc to silver was not false, for in what Volta meant by
batteries it did. In ours, of course, it flows from silver to zinc, but that does
not contradict what Volta said.

One last example may clarify the way that higher-level taxonomies
are embodied in the lexicon and also extend understanding in the sorts of
knowledge their embodiment brings with it. Having developed the example more fully elsewhere, I shall be both dogmatic and extremely selective
in calling upon it here.[14] As with electricity, I begin with the learning situation. How, I am about to ask, do students acquire the lexicon of Newtonian
mechanics, especially the terms *force, weight,* and *mass*?

A first part of the answer is that they cannot begin until a considerable antecedent vocabulary is in place, together with the ability to use it.
Unlike the case of static electricity, however, some of that vocabulary is
itself technical, acquired during earlier schooling: a mathematics vocabulary sufficient to describe trajectories, the motion of bodies along them,
and the manipulation of extensive magnitudes; some terms for the category of physical objects. In addition, though these are not strictly required,
the lexicon with which students begin ordinarily includes commonsense,

14. See Kuhn, "Possible Worlds in History of Science."

pre-Newtonian versions of the terms to be learned: *force, weight,* and *mass,* the latter perhaps labeled *quantity of matter.* In teaching Newtonian mechanics, this part of the lexicon must be restructured, a process for which there was no equivalent in the case of static electricity, where the terms to be learned were new.

Restructuring usually begins with the term *force,* and the main technique deployed is exposing the student to examples of forced and of force-free motion. For this purpose, either direct exhibit or description in the antecedent vocabulary can usually be employed. The students' previous use of *force* plays a role, but a considerable redistribution of examples is required. In the students' earlier lexicon, the standard example of a forced motion is the hurled projectile; force-free motions are exemplified by the falling stone, the rotating flywheel, or the spinning top. In the Newtonian lexicon, on the other hand, all these are examples of forced motion. The only force-free motion is one that cannot be directly exhibited and must be described with previously available terms. It is motion in a straight line with constant speed, and to declare it force-free is to state Newton's first law of motion: in the absence of an applied force, motion continues in a straight line with constant speed. That law supplies a feature of the space within which referents of the term *force* can be singled out, and it also tells something about the way that forces behave. It points in two directions: outside toward the world and inside toward the lexicon with which that world is described; it displays aspects both of the analytic and [of] the synthetic.

Newtonian mechanics includes other laws that serve such dual functions. The concept of force is quantitative as well as qualitative, and its quantitative aspect is best introduced through exposure (again, either by exhibit or in words) to the spring balance. That use, however, requires recourse to two other laws of mechanics: Hooke's law and Newton's third law. And as the lexical acquisition continues, proceeding from the term *force* to the terms *mass* and *weight,* other instruments and other laws prove to have the same dual function. The distinction between the referents of *mass* and *weight* is, for example, closely tied to the distinction between the quantities measured by the pan balance and by the spring balance. And that distinction in turn invokes the law of gravity. Here again knowledge of the world and knowledge of the terms in which the world is described are inextricably mixed, and there are still other examples.

That being the case, it is essential to emphasize once more that there is no preset list of examples or of laws to which students must be exposed in acquiring the Newtonian lexicon. Newton's first and third laws probably are essential; Hooke's law may be required as well. But elsewhere there are options. The paper from which I am excerpting bits of this example outlines three distinct routes from the acquisition of *force* to the full Newtonian lexicon, which includes the terms *mass* and *weight*. Along the first, Newton's second law is stipulated during the lexical acquisition process, as his first law had been stipulated in the establishment of the term *force*. Once the lexicon's structure has been established in that way, the terms it contains can be used to derive the law of gravity from observation. Along the second route, the law of gravity is stipulated during lexical acquisition, after which the second law can be obtained empirically. A third route fixes lexical structure by stipulating the vibration period of a known weight at the end of a spring of known elasticity. In practice students have usually been exposed to all three of these routes before they control the required lexicon, but in principle any one of the three would do alone. All three, that is, result in the same lexical structure. Individuals exposed to any one of them will agree in their identifications of forces, masses, and weights, and say the same things about them. At least within the realm of Newtonian mechanics, their communication will be unproblematic.

At last I am in a position to address the problem of apparent sharp practice to which I referred earlier in the lecture. Repeatedly today and occasionally yesterday I have switched back and forth without apparent discrimination between the material and the formal mode. Or, to put the point another way, I have appeared to conflate issues of metaphysics or ontology—what there is for terms to refer to—with issues of epistemology—how the referents of those terms are picked out. Very likely the procedure has appeared entirely circular. I shall close by arguing that it is not.

Let me underscore at the start the very large role played by the world in the lexical acquisition process and the correspondingly small role played by anything quite like definition. That is obvious in the case of terms like

star and *planet*, terms which I previously said could be learned by direct ostension and which then provided the community's closest approximation
to an observation vocabulary. But it applies equally, if less obviously, to the
acquisition of higher-level kind terms like *electricity* and *force*, terms which
are introduced contextually, within statements that are otherwise cast in
the vocabulary antecedently available.

The function of such statements—often called *contextual definitions*—
has traditionally been conceived as relating new words to old, thus providing a partial substitute for an explicit definition, such as "*bachelor* is
equivalent to *unmarried man*." But that view of them seems to me mistaken.
Statements like "An electrified body attracts chaff" function in another
way. They do relate words to words but only with the external world as an
essential intermediary. Just as one can scrutinize bodies that one is told
are stars and planets, looking for shared and for differentiating features, so
one can scrutinize the verbally described situations through which terms
like *electricity* and *conductor* are introduced. When that occurs, the image
of the situation evoked by the antecedent vocabulary plays the role that
might otherwise be played by the situation itself. Indeed, as I've already
indicated, the actual and described situations are often interchangeable.

Which brings me to my point. To the extent that acquiring the lexicon
of a language community depends on a process like ostension, the acquisition process must invoke the actual world, either by exhibiting it or by describing the way things occur in it. That is how laws and other descriptive
generalizations get involved in the lexical acquisition process. But a person
who uses the lexicon thus acquired is not bound by all the generalizations
or examples that played a role in its acquisition. That role has not made any
of those generalizations analytic.

Different individuals can, as I have said, acquire identically structured
lexicons by traversing different routes. Features that one person encounters
in the learning process may be acquired later or not at all by another. It is
only the structure of the lexicon, not the feature space in which each community member embeds it, that need be shared. Given that shared structure, each can learn things that the other knows, and they can also proceed
together to learn new things about the world. Among those new things,
furthermore, are corrections of laws and generalizations encountered

during the learning process. Some of the examples ostended during the process of lexical acquisition may prove illusory; some of the descriptive generalizations may, without precipitating a crisis, prove false. There is always some play in the system, some room for adjustment. Though one may not, for example, call into question all three of the alternate routes to the Newtonian lexicon, the structure of that lexicon could probably withstand the adjustment of one or two.

What one is committed to by a lexicon is not therefore a world but a set of possible worlds, worlds which share natural kinds and thus share an ontology. Discovering the actual world among the members of that set is what the members of scientific communities undertake to do, and what results from their efforts is the enterprise I once called normal science. But the set of worlds open to them is limited by the shared lexical structure on which communication between the community's members depends, and scientific development has sometimes had to breach those limits, to restructure some part of the lexicon and gain access to worlds that were inaccessible before. A number of such episodes separate Aristotle from Newton; a single one separates Faraday from Volta, or Planck's early readers from those who read Planck today. It is their occurrence that creates the textual anomalies from which my first lecture began, and it is these anomalies which truth-preserving translation cannot remove. When translation fails, there is other recourse, and it provides the first topic for tomorrow's lecture. But that recourse does not license the use of such terms as *true* and *false*. Their role is restricted to the evaluation of choices between worlds—worlds to which the structure of the community's lexicon gives access.

Embodying the Past

In the first lecture of this series I presented three examples of the sort of quasi-ethnographic interpretation the historian requires in order to understand a body of past belief—to make it plausible and coherent. Yesterday, in my second lecture, I suggested that the need for such interpretation arises from a disparity between the taxonomy current in the historian's society and that of the society for which the texts he studies were originally written. Sketching a model of the way in which the lexicon of natural-kind terms supplies taxonomic categories and ways of applying them, I further argued that these taxonomic disparities between past and present preclude truth-preserving translation of some central past beliefs into modern terms. That argument applied, furthermore, whether the translation was from a foreign language or from an earlier version of the translator's native tongue. The primary and indispensable role of judgments of truth and falsity, I concluded, was played within history, not across it. Terms like *true* and *false* need function only in the evaluation of the day-to-day choices made within a community that has an ontology of kinds and a corresponding lexicon in place.

My lecture also indicated, however, that barriers to the evaluation of truth claims need not be barriers to understanding, and understanding is the objective of the historian's quasi-ethnographic task. Historians and their audience need to acquire the lexicon of the community under study. They must assimilate its taxonomy of natural kinds, and otherwise situate themselves imaginatively in its world. But they need not and cannot properly employ their own lexicon and their knowledge of their own world in a piecemeal evaluation of the truth claims of that older community. Though there are judgments they can make about the past, those judgments are not properly cast in the vocabulary of *true* and *false*.

Obviously that position is deeply problematic, and four of its problems provide my topics for today. First is the problem of bridgeheads: How much communality is required to explain the success of the historian (or anthropologist) in reconstructing the belief system of another time (or another society)? Second is the problem of relativism: Can the truth or falsity of a belief about the world depend on the lexicon of the community within which that belief is held? Third is the problem of realism: Can talk of the other worlds of other communities, other cultures, be heard as anything but the wildest of metaphors? And fourth is the problem of connections between present and past or past and present: If the lexicon of a past community makes its world foreign, how can that world have become our own; how can it be our past? Those are large questions, and this is my last lecture. I must be briefer than I could wish.

1

I have suggested that past beliefs are recaptured by language learning, or rather by acquiring the lexicon of kind terms in which those beliefs were stated. That process, when successful, produces bilinguals, but it need not produce translators. The two lexicons which result might even be separately compartmentalized, with no concourse whatsoever possible between them. Where two community-based languages are involved, compartmentalization so extreme is not observed, and I shall shortly suggest reasons to suppose it never will be. As anti-relativists regularly emphasize, speakers of one human language seem able always to find a bridgehead from which to enter another, and some such bridgehead is essential to the acquisition of a second lexicon. But the required bridgehead need not be particularly broad or solid. In principle, it need not permit truth-preserving translation at all. In practice, it doubtless does permit some, but only within a restricted range.

If a bridgehead is to serve its function, then some of the taxonomic categories provided by one lexicon must overlap substantially in their membership with categories in the other. In particular, that must be the case for some categories whose members can be picked out by direct ostension, by pointing. Think of the referents of the term *star* in the lexicons of Ptolemaic and of Copernican astronomy, or the referents of *motion* in the lexicons of Aristotelian and Newtonian physics. Overlap of this sort is surely prerequisite to the acquisition of a second lexicon, but as these examples will themselves remind you, only overlap, not identity of membership, is required. The Sun is a star for us, but a planet for the Greeks; the growth of an oak was a motion for Aristotle, but not for his Newtonian successors.

In addition, there need be no particular categories in either lexicon which can, in advance of investigation, be guaranteed to overlap categories in the other. Yesterday I remarked that it was difficult to imagine a culture without a term for stars, but it's not impossible to do so. The heavens might be referred to as "spangled" (having a texture of a certain sort), without there being anything sayable about what the heavens were spangled with. When trying to acquire a second lexicon, the category stars is a likely place to look for overlap, but no overlap need be found. Lexical acquisition requires only that, for any pair of languages, overlap be found in enough places to get the learning process off the ground. Nothing remotely like a universal set of shared observational primitives is needed, and it is fruitless to look for one.

These points can usefully be put in a negative way. It is essential to the acquisition of a second lexicon that a learner be able to form and test hypotheses about the particular object or situation to which a user of that lexicon is referring when employing a particular word or phrase. And it is essential also that some of the objects or situations grouped together by the new lexicon be grouped together also by the old one. But these are conditions only on the referents of lexical items; their successful fulfillment supplies no information at all about what those items mean. With respect to meanings, the bridgehead need not supply any constraints at all. The establishment of meanings requires a second, independent process, one that I yesterday described at length when dealing with the acquisition of a first lexicon. The learner must find features shared by the various

occasions when a given term was invoked as well as features which differentiate those occasions from others when the learner anticipated the same
term but did not encounter it. No two learners need select the same features, but each must select features which yield the same taxonomy, the
same lexical structure, the same similarity-difference relations among the
referents of natural-kind terms. Otherwise they will pick out different referents for the same terms, and communications involving those terms will
break down.

Differences in lexical structure, I have been arguing, are what limit the
possibility of truth-preserving translation. The requirement that natural-
kind terms be projectible—a vehicle, that is, for inductions and for the
statement of natural laws—blocks the apparent remedy. One may not simply add terms from the new lexicon to the old, and then use the expanded
lexicon to translate. As I emphasized yesterday, natural-kind terms for categories at the same level may not overlap. If two terms share some referents, they may not both be marked as natural-kind terms in the lexicon.
That being the case it is important to emphasize that there is no reason
why a group's repertoire of features and of terms referring to them cannot
be expanded indefinitely. On the contrary, to learn new features is to learn
new ways of discriminating, and the shared biological heritage of *Homo
sapiens* makes it natural to suppose that a discrimination used by any normally equipped human being can be learned by any other.

The enrichment that results from studying the past or immersing one-
self in another culture is an enrichment of available features, available discriminations. And the universal learnability of features, which permits this
enrichment, appears to be what guarantees the existence of bridgeheads
from which differently structured lexicons—taxonomies at cross-purposes
with our own—can be acquired. The two lexicons may be everywhere differently structured; they need contain no coextensive natural-kind terms;
no statement that includes the name of a natural kind need be translatable
between them. Lexical acquisition requires only the ability to scrutinize
objects or situations named in one's own lexicon and to discern features—
often unrecognized before—which will regroup them in conformity with
the lexicon to be acquired. Sharing features, actual or potential, is a sufficient basis for the constitution of a bridgehead.

In practice, however, as yesterday's lecture indicated, bridgeheads ordinarily do comprise far more. In historical development, at least, major mismatch of lexical structure is to be expected only when comparing lexicons from very different periods or cultures—antiquity and the seventeenth century, for example—and even these mismatches are not total. In any case, they are of concern only to the historian trying to look backward across them to the point from which his narrative begins. As that narrative moves forward—within the developmental process itself—structural changes take place in smaller, isolable stages. Ordinarily, these are restricted to one or another local region of the lexicon: the region containing *force*, *mass*, and *weight*, for example, or *battery*, *resistance*, and *current*, or *oscillator* and *energy element*. Outside these regions, the structures of the altered and unaltered lexicons are homologous, and truth-preserving translation is unproblematic. The bridgehead in actual historical transitions is strong.

Nevertheless, it is only a bridgehead. There remain those regions of the lexicon in which structure differs, and these regions contain the fundamental terms of a science, terms which, like those just mentioned, are required to state its constitutive generalizations and laws. Like other statements containing those terms, these generalizations are not translatable. Modern knowledge cannot be used to judge them true or false. Truth values applicable to them can be supplied only from within the lexicon used to state them.

2

My insistence that many statements constitutive of a science can be assigned truth values only from within the community of practitioners has made my position seem relativistic, and perhaps it is. The epithet *relativist* is used in so many different ways—its referent is identified by so many different features—that discussion of the question relativist-or-not is unlikely to prove fruitful. Let me substitute another for it: Has anything worth preserving been lost?

I think it has not, for what I am relativizing is not truth value but effability. If the same statement can be made with the lexicon of different communities, it must have the same truth value in all. My point has rather been that some statements which are clear candidates for truth value in one community are simply unsayable in another. The situations they describe do not occur in any of the possible worlds to which that community's lexicon gives access. It is these statements I have been describing as impossible to translate with the specificity required by the truth-value game.

Using the techniques of ad hoc compromise employed by actual translators, one can approximate the content of such statements, and it is these compromise statements, made with one's own lexicon, that one often feels obliged to judge. But there is nothing to be achieved by doing so. Translations of that sort are not usually truth preserving. Asked to say whether some such compromise translation is true or false, the historian who has acquired the lexicon in which the original was stated is often perplexed how to respond. Consider once again the statement "The Sun is a planet," intended to record something the Greeks believed about the heavens. Using our lexicon, there is no better way to render the Greek belief, and with our lexicon that statement is plainly false. But what has that got to do with what the Greeks believed? Given the taxonomy of celestial kinds embodied in the structure of the Greek lexicon, the Sun was a planet. To render their belief in our words as "The Sun is a planet" is a misleading compromise. No better translation is available, but the truth value that should attach to this one is at best unclear. Does it really make a difference whether the question of its truth has an answer or not?

To see the perplexity in a more typical form, consider the question I asked at the end of the first lecture: When Aristotle said there could be no void in nature, was he simply mistaken? Was his statement false? If the terms in that statement are ours, the answer is surely yes: the statement is false. We know there can be empty space, place without matter. Already in the seventeenth century the barometer and the air pump provided convincing evidence of that. But Aristotle, you will remember, did not mean the same thing by *matter* as the new, corpuscular philosophers of the seventeenth century. For him, matter was a neutral substrate, everywhere available to receive form and by doing so to constitute a substance, a body, a thing. Where there was no matter there could not be body, not even potentially. Using terms from the Aristotelian lexicon, the very conception of place without matter was incoherent. Having learned what the terms mean,

acquired the requisite parts of Aristotle's lexicon, and experienced its coherence, do we still want to say that Aristotle's statement was false?

I do not. But neither do I wish to declare the statement true. Rather, I
want to hold off the question, distance myself from it, inquire what exactly
I am being asked. I cannot recast Aristotle's statement in my lexicon and
then label it true or false as I would label a statement of my own or of a
contemporary. But if I instead adopt Aristotle's lexicon and attempt to answer in it, I suffer the benefits of hindsight. I know the arguments pro and
con. (They do not include the barometer or the air pump, which are only
indirectly relevant.) And I also know something about the way the balance
between them shifted with time. But those arguments, each in its time, always left the issue in doubt, never permitted a decisive choice between true
and false. Their effect, together with that of many other events and issues,
was to create pressure on the lexicon, pressure that ultimately resulted in a
revision of its structure, an altered taxonomy, and a changed set of natural
kinds for discussions of bodies and their properties, of space, and of motion. But that outcome was not falsification of the Aristotelian statement.
Instead certain key terms used in making and discussing that statement lost
their meaning. And there was no way, using the terms that replaced them,
to restate what it was the Aristotelians had in mind.

None of these arguments is meant to place the Aristotelian lexicon
beyond judgment. Doubtless, its seventeenth-century successor was a far
more powerful instrument for solving the puzzles with which scientists
are characteristically concerned. The adjustment of lexical structure which
made the Sun a star, the Earth a planet, and the Moon a satellite permitted more precise solutions than any available before to a larger range of
problems about celestial phenomena. And the post-Galilean taxonomy of
matter, space, and properties had the same effect on problems about the
newly restricted phenomena of motion. But the evaluation of a lexicon
with which to describe phenomena and to build theories about them is a
very different undertaking from the assignment of truth values for the individual statements that the lexicon permits one to construct. And the latter
enterprise, the assignment of truth values for individual statements, can
occur only after a lexicon is in place. Whether the statement is "Neutrinos
have no mass," "The Moon goes around the Earth," or just "It's raining," the

assignment of a truth value requires discussion of evidence and can occur only among those who know what the statement means.

Evaluating a lexicon is a different matter, for a lexicon cannot properly be labeled true or false. Its structure, the taxonomy it provides, is a matter of social or linguistic fact, like the Greek use of *motion* and *star*, like Volta's use of *battery* and *resistance*, or like Planck's use of *energy element*. Nor can a lexicon properly be described as "confused," though its use may on occasions result in confusion. Instead, one lexicon is a better or worse instrument than another for achieving specifiable social goals, and the choice between lexicons—or, better, the direction of lexical evolution—necessarily depends on those goals.[1]

With respect to lexicons, I am thus suggesting, the pragmatists were generally right. Lexicons are instruments to be judged by their comparative effectiveness in promoting the ends for which they are put to use. The "choice" between them is interest-relative. With respect to lexicons my position is instrumental and relativistic. But relativism with respect to lexicons need not bring with it relativism with respect to truth, and I think it vital that it not be allowed to do so. To the extent that the members of a society are bound together from hour to hour and day to day, it is the truth-value game—most centrally, the law of noncontradiction—that provides the ties between them. Where that law applies, differences are discussable and agreement on the basis of evidence can be anticipated. Where interests enter, the fragmentation into communities begins, discussion becomes problematic, and agreement on the basis of evidence is at risk. But if the truth-value game does promote solidarity, then about truth the pragmatists must be wrong. Truth cannot be warranted assertability: two members of a community may with warrant assert contraries, but it is a rule of the game that only one of them can be right; the dissolution of community starts with the violation of that rule. Nor can truth be the ultimate limiting product of the process of rational inquiry: as a central requisite of discourse and negotiation, it is required from day to day; to place it in principle beyond

---

1. Theories are somehow in an intermediate position. They help to determine a lexicon and are thus to be judged as instruments. But they also determine individual statements which are to be judged as true or false.

current reach is to block its function. Human communities, I am suggesting, are discourse communities, and the truth-value game is essential to them. That is most obvious—both in the breach and the observance—for communities of scientists, but I take its applicability to be universal.

3

Though the position just outlined presents difficulties, relativism seems to me an inappropriate name for them. Truth claims and their evaluation remain in place where they are needed, within a community's day-to-day life. Between communities distinguished by lexical structure, the evaluation of truth claims is often impossible. But there are then bridgeheads available sufficient to permit judgments about the way of life (or the way of practicing a science) that the other community's lexicon permits. If that way of life seems superior with respect to what one values (if, for example, it permits the solution of some previously unsolved technical problems), then one can migrate to the other community, acquire its lexical structure, and thereafter go native, abandoning the lexicon with which one was raised. Metaphorically at least, that is the way much scientific progress occurs. And the reality of progress is not for me in doubt.

But that progress is instrumental. Though real, it is not progress toward reality. Science does provide us with a more and more powerful taxonomy for dealing with the world, but it does not do so by discovering a lexicon-independent truth. We cannot properly say that the Greeks were wrong to identify the Sun as a planet or that Aristotle was mistaken about the impossibility of a void. Nor may we say that Volta mistook the direction of current flow in a battery, or that Newton was wrong to suppose that the simultaneity of two events was independent of the coordinate system from which they were viewed. In its time, each of these beliefs was stated or derived from statements that called upon parts of the lexicon differently structured from our own. Those statements are not translatable; they cannot individually be compared with statements we might make; our judgments of truth value cannot be applied to them.

Nor can we say with respect to these statements and others like them that, though not quite right, they were approximations to the truth toward which the sciences move ever closer. When comparison is impossible, what can be the meaning of phrases like "closer and closer to the truth"? Besides, all such phrases—"zeroing in," "cutting nature closer to its joints," and so on—imply that one statement can be truer than another, and are correspondingly irreconcilable with the law of noncontradiction. Truth and falsity do not admit of degrees. To permit them to do so is, I've suggested, to abandon a fundamental requisite of discourse, of negotiation, and of the community they underwrite. Moreover, with respect to natural kinds, to the ontology embodied in the lexicon, the history of science displays nothing like zeroing in. While problem solutions advance steadily in number and precision, the ontologies from which those solutions derive vary widely in numerous directions. To date no one has shown how to display anything like an asymptote toward which science, throughout its history, has been moving closer. At any time, there are best scientific guesses about the nature of the world's ultimate parts, but they do not stay in place as lexicons change.

Considerations like these have in the past led me to speak of the world's changing during the revolutionary episodes here glossed as times of change in lexical structure. Alternatively, I have sometimes said that, after a revolution, scientists live and work in a different world. These lectures have perhaps suggested why such metaphors seemed appropriate. But to talk of the world's changing is also misleading, and I have been among those misled. Let me therefore try, however tentatively, to point the direction in which a more literal and defensible position may be found.

The difficulty with suggesting that the world changes is that it invites questions like "Do you mean that there were witches in the seventeenth century?" or "Was there phlogiston in the world of eighteenth-century chemists?" To such questions I have sometimes answered, yes, but always in a most equivocal and embarrassed tone. Now I see that what I should have done is reject the questions as ill formed. What proves misleading about saying "the world changes" is the implication that the community stood still while the world changed about it, so that a statement which was true before the change was false after it. In fact, both the world and the community changed together with the change in the lexicon through which they interacted.

Consider the term *phlogiston*. During the central third of the eighteenth century it was for many chemists a natural-kind term in a lexicon which related it structurally to kinds denoted by such terms as *principle* and *element*. These terms have either vanished from or changed their position in the lexicon of modern chemistry, and the structure of the new lexicon offers no place to fit them in. The chemical principles, for example, had been quality-bearing natural kinds, and phlogiston had been one of them.[2] If the terms that referred to them were still natural-kind terms, then many kind terms in the modern lexicon could not be. Only one of the two inter-related sets could be projectible, [and could] support inductions. Given the choice between them, there is no doubt which would be chosen by a person whose goal was solving chemical problems. For that, objective modern chemistry is the more powerful tool. But before that choice was available, the lexicon that contained *phlogiston, principle,* and the older use of *element* did successfully support inductions, and the projectible terms in it were natural-kind terms in the same way that their successors now are. In whatever sense the world of isotopes, molecular orbitals, and polymers is real, so in its day was the world of phlogiston and the other quality-bearing principles.

Not only were both worlds real, but they were in almost every standard sense also objective and external. Terms which denoted natural kinds in either of them could be learned through ostension, initially by pointing directly to their referents out there in the world and subsequently by describing, in an antecedent shared vocabulary, situations in which their referents occurred. The world described with the resulting lexicon (either the earlier or the later one) was solid. Theories could be developed about the natural kinds it contained, and the truth values of statements that followed from those theories could be assigned by observation and experiment. Agreement about those assignments was ordinarily to be expected, and in case of disagreement a request for explanation was always legitimate. Though the individual community members involved in these activities might differ subjectively—in their tastes, for example, or their interests—those differences would seldom affect their assignments of truth values and would

2. For more on this subject see my "Commensurability, Comparability, Communicability," in *PSA 1982*, vol. 2, ed. Peter Asquith and Thomas Nickles (East Lansing, MI: Philosophy of Science Association, 1983), 669–88 [reprinted as chap. 2 in *Road Since Structure*].

never override the obligation to give reasons for disagreement. The world inhabited by community members was, in short, intersubjective.

But intersubjectivity, like the ability to establish shared truth values, need extend only to the boundaries of community. Among the usual criteria of reality and objectivity, mind independence is still missing, and I think it will not be regained. Minds are a part of what constitutes a world, and there can be no worlds without them. I do not think of the mind of one or another individual, for a world is not subjective. Nor do I think of the mind of a group, for a group does not have a mind. But there is no world without a group of living individuals who share it, who interact with it in reliably similar ways, and who interact with each other in ways that presuppose it. For the higher animals, at least, what the actions of individuals presuppose is a shared way of dividing the world into interrelated kinds. And for humans, those kinds have names which are found in and interrelated by the mental structures I have been calling lexicons.

By now it will be clear that the position toward which I grope is vaguely Kantian. Categories of the mind are required to constitute experience of the world; without them there is no experience. But if the position is Kantian, as I suppose, it is Kantian with two differences. First, the categories which I seek are not those of a human individual nor are they common to all human individuals. Their locus is rather a historically situated community, a group whose members share a lexical structure acquired from their parents and teachers and who in turn transmit that lexicon, perhaps in altered form, to their successors. Second, as my reference to alteration indicates, these categories can vary—not totally but in considerable measure—from group to group and also, for a single group, with the passage of time.

Communities cannot, of course, make up whatever categories [their] members please. The world described by a lexicon is solid. [The world] participates in the determination of truth values, in distinguishing true statements about it from false. The lexicon is a product of human minds at work, not at play, and there must be something for them to work on. The same something guarantees the existence of bridgeheads from which the lexicons of other communities can be explored. But about that something we cannot speak. Like Kant's *Ding an sich*, it is ineffable, prior both to worlds and to the communities which inhabit them. Speech requires not only something to talk about but also someone to talk and someone to listen. With all three in place, there is already a world and already a speech community. Both are historically situated, in interaction through the lexicon which is constitutive of both. About what came before, there is nothing that can be said. Not even the question, "Which came first, the world or its group?" is admissible.

4

I am left with two questions, both about historical development. First, given the role of a lexical structure in constituting a world, how can a lexicon be changed? Second, given the problems of translatability that result from lexical change, what connections with the past are available to the present, [and] how can the past be part of present identity? The first question has already been touched upon. Aspects of a community's knowledge of the world are built into the structure of its lexicon, and novel experiences sometimes strain that built-in knowledge in ways that can be relieved only by lexical change. Such strains can come about in various ways, but a single simple example will show why I use such terms as *strain*.[3] For that purpose, I return for the last time to Aristotelian physics, in particular to the Aristotelian concept of motion.

In the first of these lectures I said that the Aristotelian term translated as "motion" refers to changes of all sorts: from sickness to health, or [from] acorn to oak, as well from place to place. Motion is thus change of state, and its salient features are its two end points and the time that elapses in the transition between them. But motions, or at least local motions, have another salient feature, one which I shall call *perceptual blurriness*. Aristotle invokes it from time to time in discussing a motion's speed. Ordinarily, as one might expect, he takes speed to be directly proportional to the distance between the end points of a motion and inversely proportional to the time required for the motion to take place, a conception that evolves into our

---

3. Cf. "The Function for Thought Experiments," most conveniently available in my book of essays, *The Essential Tension: Selected Studies in Scientific Tradition and Change* (Chicago: University of Chicago Press, 1977), 240–65.

notion of average velocity. On other occasions, however, he speaks of the speed's increasing or decreasing during the course of a motion, a conception that evolves into our instantaneous velocity. For Aristotle, however, these are not two concepts or sorts of speed, but two aspects or features— one static, the other dynamic—of a single concept.

Both Aristotle and his scholastic successors regularly acted as though these two features fit smoothly together, as though any motion which was faster when judged as a whole would also be faster during each of its parts. That was part of the knowledge of nature embedded in their lexicon. Under most circumstances it presents no problems, and in an only slightly different world—one in which, for example, all motions occurred at uniform speed—it would present none at all. But there are situations in which the two conflict: a motion judged faster by one criterion is slower when judged by the other. Galileo exploits one such situation unmercifully in the first [day] of his *Dialogue on the Two [Chief] World Systems*, an important episode in the lexical restructuring which made motion a state and speed a characteristic not of whole motions but of motion at an instant.[4]

In the *Dialogue*, Galileo asks his interlocutors to imagine two inclined planes, one vertical, or very nearly so, the other oblique. (See the diagram below.) Two balls are to be released simultaneously from the tops of these inclines, and Galileo asks which motion will be faster. The first response is unanimous: the motion along the vertical plane is faster; the direct dynamic perception of speed has prevailed. But Galileo now reminds the other participants in the discussion that the two planes are of very unequal length. To compensate for that difference, he marks off from the bottom of the oblique plane a distance equal to the length of the vertical plane and suggests that it be used in judging the speed of the oblique motion. When he asks again which motion is faster, the initial answer is reversed. Motion along the oblique is declared faster, for the same distance is there covered in less time. This time, however, the tone of the response gives evidence of

---

4. Other important episodes contributed, many of them much earlier. Among them were the introduction of a distinction between the latitude and longitude of a motion, and also a recurring debate about whether motion is a *fluxus formae* or a *forma fluens* [a flowing of successive forms or a single form that flows]. Both these developments were scholastic, centering in the fourteenth century.

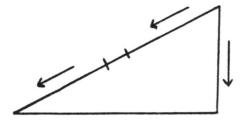

uncertainty, discomfort, and constraint. After that Galileo moves the segment he has marked off on the oblique plane to the top of that plane, to the middle, and again to the bottom, asking in each case whether the motion along the oblique or along the vertical is faster. Answers are inconsistent and they cease to be unanimous. Finally the participants—tugged repeatedly in opposite ways—realize or are shown that their difficulties are conceptual. Long-standard use of terms like *motion, speed,* and *faster* are not well suited to describe certain aspects of phenomena that they see every day.

Reading that passage from Galileo, I am regularly reminded of an episode from my school days which a few of you may share. Einstein, I was then told, had discovered the relativity of simultaneity. I did not doubt that attribution, but neither could I understand it. To me it seemed vaguely ungrammatical. Surely, simultaneity was like truth: it could not be relative without losing its meaning and its force. The relativity of simultaneity was not the sort of effect one could discover. What was it that I was being told? Later I found out—through exposure to the famous thought experiment about the moving train struck by lightning at both ends. When an observer outside the train reports that both ends were struck simultaneously, an inside observer reports that the front of the train was struck first. But what I learned from that experiment was not simply a fact about simultaneity, but equally and more fundamentally [a fact] about space and time, about clocks and metersticks. And these lessons were not simply factual, though they were that. Equally, they were conceptual and lexical; I learned to think differently about what space and time were and to make different use of the vocabulary that referred to them. Being told that Einstein had discovered the relativity of simultaneity helped to prepare me for that lexical change, but I do not think I was wrong to find it ungrammatical.

These examples are not typical. There are, as I said in introducing the
first of them, other routes to lexical change. But these have the special virtue of displaying clearly the lexicon's involvement in revolutionary change
of theory. By doing so they locate the strain that precedes such change
where it belongs—on the lexicon.

5

The remaining question is the one that, months ago—before I discovered
how long it would take me to reach it—gave this third and last lecture its
title. How can historical narratives bridge the gaps or ruptures that remain
after lexical change? How, to put the point differently, can the past be trans-
ported to the present and incorporated in[to] present identity? I think
those questions have two answers, both inescapable, and mutually deeply
inconsistent. I shall close these lectures with a few words about each and
about their relationship.

The first answer presupposes the sort of history I have described repeatedly in these lectures. A historical narrative opens with an ethnographic
or hermeneutic reconstruction of the relevant aspects of some period in
the past. It sets the stage and introduces the actors, their lexicon, and their
world. After that the story proper begins, always narrated in the specious
present, always viewed on a synchronic screen. Strains on the lexicon are
described as they arise; the resulting confusions are displayed; the first
metaphorical attempts at lexical adjustment are introduced; and the episode culminates with the establishment of a new lexicon, a new life-form,
a new world. A single narrative may include one or a long succession of
such episodes. It may close at some time in the past or at a time close to
the present.

That is the sort of history that I have tried to promote, both by producing it myself and by teaching others to do so. I believe in it deeply. To the
extent that it succeeds, it makes possible new discriminations, opens doors
to new worlds. In this respect, it is like any successful ethnography. But because it deals with a moving present, it does something that ethnography
cannot. It displays historical evolution at work, always moving blindly into
a still nonexistent future. It exhibits also something about the nature of the

products of that evolution, one of which is human knowledge. Carried out in this way, history really is philosophy teaching by example.

But there is also an important function that this sort of history cannot serve. We can study it and learn from it, but always as the story of an alien tribe. The story of our past, the one that has made us what we are, must be written differently, using our own lexicon. In it, the Sun was always a star, motion was always a state governed by Newton's first law, the power of the battery was always chemical, simultaneity was always relative, and so on. On all these subjects, people once held erroneous beliefs, corrected during the course of history through brave struggles against ignorance or bigotry. We can, this view holds, explain why these beliefs were held, but they were false nonetheless.[5] This sort of history does not attempt to acquire and deploy the older lexicons used by the communities under study; instead, it tells the story of the past in translation, abandoning the attempt to preserve truth values in exchange for preserving a constant truth. The resulting narrative does not proceed blindly but is constantly guided by what the translator—already possessed of and by the future—thinks it important to achieve. Widely known under the label of *Whig history*, this narrative mode coincides with the more ethnographic approach previously described only during the period since the last alteration of lexical structure, a period for which it is debatable whether history can be written at all.

If one's concern is, like mine, primarily philosophical, directed to the nature of the historical process or the nature of human knowledge, there is no question which of these modes is to be preferred. I am well known for the consistency with which I have denounced and ridiculed Whig history. Given my concerns, I do not mean to quit. But it is correspondingly important for me to insist that Whig history has an indispensable human

5. In principle, these explanations of past errors might follow the path I have taken in these lectures, acknowledging older lexical structures but insisting that those structures mistook the nature of the world and were correspondingly false. The argument between that approach and the one I have been developing would then show a distinct resemblance to the historical debate over special relativity: its opponents insisting that there really was a preferred frame and that the contractions of moved rods were real physical contractions, etc.; its advocates insisting that, in the absence of a way of locating a preferred frame, it was more reasonable to abandon the concept, suppose all inertial frames on a par, and alter our conceptions of space and time accordingly.

function that cannot be fulfilled by the sort of history in which I most believe. It supplies community members with a past which is not foreign but
domestic, which can be assimilated directly, and which can serve as a platform from which to move ahead.

To illustrate the cleavage I have in mind, let me speak briefly from my
experience as a one-time scientist turned to history of science. In the latter
role, I was often asked whether science ought not to be taught historically,
and I regularly responded that it should not, at least not to students who
hoped later to practice science.[6] These students need to acquire command
of the current tools of the profession, both conceptual and instrumental;
history is at best a slow and inefficient way to do that. Learning alternate
possibilities—how Aristotle conceived motion or the void, how Volta conceived the battery, how Planck conceived his first derivation of the black-
body law—these would be irrelevant distractions. Conceivably, it might
even undermine the student's grasp of and faith in the tools required for
current practice. Though the sort of history I have practiced should, I feel,
be taught to a variety of audiences, including scientists, it is not an appropriate route to professionalization. Future professionals should learn their
tools directly.

Somewhere here there lurks an aporia. The tools that a professional
requires are products of history. And those who use them must see themselves as full-fledged participants in the historical process by which those
tools were and will be developed. Like everyone else, their present identity requires an appropriate past, and that is provided by a sort of narrative
which, at its not infrequent worst, simply attributes bits of current knowledge, items from current textbooks, to historical figures who are supposed
to have discovered them, displacing ignorance or superstition in the process. The very inadequacies, factual and conceptual, of that approach make
it a more effective source of the identity required for the successful practice
of science. If they have been formed by a lie—which I do quite intend to
suggest—it is a noble one.

6. What I was counseling against was a systematic use of history in presenting the subject matter the student would need. One or two detailed case histories of individual scientific
advances might very well be useful to prospective scientists, but only after they had already
learned the science involved.

I exaggerate, of course, but in these closing moments I could not do otherwise. The two narrative modes I have been describing are never found in pure form; to a greater or lesser extent, they always interpenetrate. But the cleavage between them is nonetheless real. In a more muted form, for example, it separates historians whose subject is their native land and those who study other nations or other cultures. It is the cleavage between those who need history to look back and those who need it to look ahead, and it will not be eliminated. Though it has emerged explicitly only in my closing remarks, concern with it has run through these lectures. Seeing no way to discharge that concern, I shall leave you with it.

The Plurality of Worlds
An Evolutionary Theory of
Scientific Development

Abstract

Preface

This book is a return to the central claims of *The Structure of Scientific Revolutions* and the problems that it raised but did not resolve. Both works argue that it is incoherent to think of scientists as trying to discover objective truths about the real world. Recognizing the nature of this incoherence opens the way to a reassertion of the cognitive authority of the sciences.[a]

Part I: The Problem

Chapter 1: Scientific Knowledge as Historical Product

The view of science as a historically situated, changing practice is the starting point of this work. A brief overview of the projected book then follows, chapter by chapter.

Section I. The traditional and the developmental philosophy of science have very different ways of producing historical accounts of past science, and of putting them to philosophical use. This is shown by contrasting a traditional with a developmental account of Torricelli's discovery that nature does not abhor a vacuum.

Section II. The traditional philosophy of science was plagued by two problems: the need for a neutral observational vocabulary and the fact that no test result could be conclusive. The developmental approach to science has no need for a neutral object language, nor for an account of conclusive testing of an isolated hypothesis.

Section III. Many criticisms of the developmental approach stem from the strange way in which philosophers employ the terms *objective* and *subjective, rational* and *irrational.* The concepts which these terms denote need systematic philosophical scrutiny. Methodological relativism concerns rationality, and is neither new nor a problem. The real problem is not about truth itself either, but only about truth as correspondence.

Chapter 2: Breaking into the Past

Scientific knowledge is a product of a particular developmental process; to understand it, philosophy of science must rely on a hermeneutic, ethnographic history of science. The best way to show this is through particular examples.

Section I. If we understand Aristotle's physics as an integrated whole, with concepts different from ours, we will understand why Aristotle *had* to think that void is impossible.

Section II. Volta's early diagrams of the electric battery seem erroneous when seen through the spectacles provided by a later physics, but make perfect sense once we recover the meanings of key terms current at the time of Volta's writings.

Section III. Planck's early work on the black-body problem should not be read from the standpoint of developed quantum theory; we need to understand that Planck's terms attach to nature differently than our terms.

Section IV. In all three examples, a historian started with an impression that the text contains absurdities, and removed them by recovering the lexical structure and the system of beliefs, different from the historian's own, that the text assumed. Incommensurability primarily obtains between historically distant ways of doing science, and is thus mainly a problem for a historian. Incommensurability between contemporaries' lexical structures is always only partial and, although it makes communication difficult, it does not prevent it. Contemporaries typically manage to disagree about substantive matters, rather than just talk past each other.

Section V. A deeper analysis of the example presented in section I of this chapter shows that Aristotle's concepts of *void, motion,* and *matter* are closely interrelated, but very different from our own. We cannot thus simply evaluate as true or false his statement that void is impossible.

Chapter 3: Taxonomy and Incommensurability

The roots of kind concepts are prelinguistic: human beings share with other animals the ability to discriminate kinds.

Section I. There are two ways of responding to problematization of the distinctions between analytic and synthetic, and between intension and extension. Quine abandons the notion of meaning or intension in order to preserve a neutral, culture-free foundation for knowledge. This book takes the other fork: sciences have localized, historically situated, and moveable foundations. Making a case for that position requires the rehabilitation of a conception of meaning that does not proceed via articulation of the necessary and sufficient conditions, but includes more than extension.

Section II. The case histories in the preceding chapter provide two important clues for the reformulation of the concept of meaning. First, the terms that require reinterpretation are all of them kind terms. Second, kind terms are regularly interrelated in localized groups that must change together if a coherent reading is to be achieved.

Kind terms are of two sorts: taxonomic kinds and singletons. The meaning of a taxonomic kind term is bound up with the meanings of the other kind terms in the same set; none has meaning independent of the others. Singletons are fundamental categories of thought: space, time, and individuated physical body; perhaps also concepts of cause, of self, and of the other. Singletons are not grouped with similar kinds in contrast sets: they are sui generis. Nevertheless, they are also interdependent and must be learned together in small local groups. In [the] natural sciences, where they play a central role, singletons must be acquired together with universal, lawlike generalizations.

Section III. Both singleton kinds and taxonomic kinds are subject to the *no-overlap principle*. For taxonomic kinds, this means that kinds within a single contrast set share no member. For singletons, the principle is stronger and amounts to a form of the principle of noncontradiction. The no-overlap principle applies to both world and language—each is an inescapable consequence of the other, and neither has ontological priority. This does not mean that there is no distinction between words and things.

Members of linguistic communities share a structured kind set, of which the roots are innate, but which is largely acquired through learning. The structure of its kind set encodes the ontology of a community and greatly restricts what community members' beliefs could be.

Section IV. The techniques and experiences of a historian struggling to break into a text are close to those Quine attributes to a *radical translator*. But Quine still seeks a fixed Archimedean platform for translation, and hopes to find a privileged class of observation sentences which can be evaluated on the basis of sensory stimulation alone. This foundationalist, empiricist, transcultural premise is mistaken.

Section V. Incommensurability as an obstacle to understanding is surmounted by learning a new language. Bilingualism precedes translation: although bilinguals can understand both languages and respond to what is said in each, what they hear and what they say cannot always be expressed in both languages. They must therefore constantly be aware in which language community they are participating. The shared biological and environmental heritage of human beings makes bilingualism possible. If we discovered a group, imputed a language to it, but could not learn that language, we would *not* conclude that we have discovered an inaccessible human language.

Part II: A World of Kinds

The goal of this part is to provide an empirically based foundation for a theory of meaning of kind terms.

Chapter 4: Biological Prerequisites to Linguistic Description: Track and Situations

Developmental studies of human cognitive apparatus provide a basis for explaining language acquisition, incommensurability between languages, and understanding across incommensurabilities.

Section I. [A] survey of recent research shows that infants are equipped with a neurologically embedded, primitive form of *object concept*: the object as a bounded region, all of whose parts move together. Tracking response in infants implies no cognitive separation of what, for us, are the concepts of space, time, and object. These concepts become differentiated only with language acquisition.

Section II. The basic *concept of a kind* is also in evidence within hours of birth. Concepts of kind and object thus have biological roots; they are widespread throughout the animal world.

Section III. Research in the field of categorical perception suggests that recognition of objects and kinds does not require knowledge of features common to all presentations of the object or shared by all members of the kind. Recognition is a noninferential process that proceeds from *perception of differentiae* rather than from perception of shared features. Differentiae supply the fastest and surest means to discriminate situations requiring different behavioral responses.

Section IV. Empirical evidence and evolutionary considerations support the claim that language plays a central role in the conceptual development from proto-objects to objects with permanence.

Section V. This chapter extends interpretation of behavior, discussed in part I, to interpretation of prelinguistic behavior. This extension presents new difficulties, since the possibility of an adult interpretation of a prelinguistic infant is extremely limited. We can still establish that the neural apparatus developed for prelinguistic life constrains what can coherently be put into words in any language at all.

Chapter 5: Natural Kinds: How Their Names Mean

This chapter begins to develop a theory of kinds which sees kind membership as established by differentiae and [the] no-overlap principle rather than by characteristic features.

Section I. The system of kinds of living organisms encountered in everyday life operates in the same way as the kinds of the natural sciences, except that the conditions required for adequate function of everyday kinds are vastly less stringent.

Section II. Reidentification of an individual is possible only when its kind is known. The identification of kinds is made possible by placing them in a hierarchy.

The properties of members of natural kinds can be established by direct observation, but which properties are in fact observed will be deeply influenced by interest and belief. However, a language community requires that its members be able to achieve ultimate agreement about the observed properties of objects.

Section III. Neither a finite set of observations nor their logical consequences can determine all the properties shared by members of a natural kind. Members of natural kinds are inexhaustible, and none of their features is necessary.

Taxonomic categories are culture-bound: learning to classify individuals into kinds involves learning the categories of a culture. Competent speakers share a lexical structure: they all cluster objects in the same way.

People raised in different cultures sometimes differ in featural vocabularies, because their cultures cluster objects into different kinds. But incommensurability, as it is experienced in practice, is always a local phenomenon. Members of one culture can enrich their featural vocabulary with the features deployed by a different culture without jeopardy to their own. Between any pair of cultures, many kinds and many elements of the featural vocabularies must be shared.

Section IV. The constraints on a kind set are pragmatic. The only questions relevant when evaluating such a set concern its success in meeting its users' needs, including their need for shared observation. Needs vary, however, from culture to culture, as well as between the various subcultures of complex societies.

An anomalous object that equally resembles members of two distinct kinds threatens the accepted taxonomy; the likely solution is a lexical redesign, but there are numerous ways of making it. A group of specialists evolves to take responsibility for such tasks, involving discovery of important similarities and differences. The society's need for intelligent answers to such questions gives the group of specialists its authority.

Section V. Materials share three prominent characteristics with organisms: the role of differentiae in their identification; the role of hierarchy in locating the appropriate set of differentiae; and the role of observations about whose outcome members of the community must ordinarily agree. Four interrelated differences are as noteworthy as the parallels. First, materials are not objects. Second, the hierarchy for materials bottoms out not in individuals which belong to kinds, but in kinds themselves. Third, kinds of materials do not change over time. Finally, the hierarchy of natural kinds of materials is much simpler than that of organisms.

Section VI. If members of two cultures (or two periods in the development of a single culture) have incommensurable kind sets, direct translation between them will be impossible. If one cannot state two competing beliefs in the same language, then one cannot compare them directly with observational evidence. That should not suggest that there are no good reasons why, over time, only one of them survives. Nor should it suggest that those reasons do not rest on observation. But it should suggest that the standard conception of a *choice* between the two on the basis of observational evidence cannot be right. Comparison requires simultaneous access to the things being compared, and that is here barred by the no-overlap principle.

Chapter 6: Practices, Theories, and Artefactual Kinds

Just as scientific taxonomic kinds arise from the natural kinds of everyday life, so the abstract kinds appearing in scientific theories arise from everyday artefactual kinds.

Section I. The nature of artefacts is dual: as physical objects they display observable properties, but it is their function that groups them into kinds. Artefacts and their functions are thus nodes in a practice, and the nodes are differentiated by relating their functions to those of other nodes which serve other functions. For artefactual kinds, there can be no talk of empty perceptual space, or of nature's joints.

Section II. Physics originates in the study of matter in motion. Both the concept of matter and that of motion are abstracted from the study of natural and artefactual kinds. But what is abstracted as *matter* and *motion* differs. Neither Aristotle's nor Newton's way of abstracting is properly described as correct or incorrect, true or false. What differs about them is their effectiveness as tools for practice in two quite different historical situations. That they are tools and that they came into existence through human agency is what makes it appropriate to group them with artefacts.*

More on artefacts. Since different artefacts belong to the same kind when they share the same function, objects of vastly different appearance can belong to the same artefactual kind. Also, a given artefact can be used in different practices; in an emergency, for example, an artefact can be used for a completely novel purpose.

Members of natural kinds can at the same time be members of artefactual kinds. For example, a member of the natural kind *dog* can be a member of the artefactual kind *rescuer* or *hunter* if it has been trained to play a specific role in a human practice.

Some artefacts are observable objects (for example: knives, saws, rescue dogs, microscopes), but other artefacts are invented mental

---

* Here end the summaries of the extant text of *Plurality*. What follows in this abstract are editorial reconstructions of the main ideas for the planned but unwritten parts of the book.

constructs which cannot be observed at all. Abstract artefactual kinds are learned through their relation to other mental constructs within a practice.

Two kinds of kinds. There are two sorts of kind concepts: taxonomic kinds and singleton kinds. Taxonomic kinds come in contrast sets within a hierarchy, but singletons do not. Both are governed by the no-overlap principle.[b]

Both types of kind terms originate in primitive protoconcepts that do not require language, and are usually exhibited by nonlinguistic animals and prelinguistic humans. Such hardwired singletons include *object*, *time*, and *space*. Cognitive modules that make possible protoconcepts also provide the basis for forming concepts (taxonomic and singleton) used in ordinary language and in science.

Unlike taxonomic kinds, scientific singletons (such as *mass* and *force*) are never directly observable, and in that respect the distinction between taxonomic kinds and singletons in science resembles the old distinction between observable and theoretical terms. In a mature natural science, singletons are usually introduced together with one or more of the universal generalizations (laws of nature); frequently, instruments are also introduced with them. However, it is possible to learn how to use singleton terms without knowing the theories that have led to their being recognized by scientists. This indicates the need for emphasis on scientific practice.

Part III: Reconstructing the World

Part III returns to the themes of part I, for which part II attempted a foundation.

Chapter 7: Looking Backward and Moving Forward

Conceptual change happens in all languages, natural as well as specialized.[c] Newly structured kind sets arise from the need to accommodate objects and processes that the old kind-set structure was not able to handle. The historical view of conceptual change finds it to be gradual, locally holistic, and in some ways showing a pattern similar to the pattern of biological speciation.

Scientific structured lexicons develop by refinement or reform of ordinary language's natural kinds. This process involves looking for regularities governing these kinds and abstracting some of their properties (for example, geometrical, logical, dynamic, and so on).

We need to distinguish two questions, which were unfortunately not properly separated in *Structure*. The first question is: How do the proponents of different lexical structures and different ways of doing science (different *paradigms*, in the vocabulary used in *Structure*) communicate with each other across incommensurability during the periods of extraordinary science? The second question is: How does a historian recover past meanings and beliefs that are incommensurable with the historian's own yet necessary for an understanding of past science?

Scientists who are one another's contemporaries share many concepts, beliefs, values, and methods; at any given time, their disagreement is local, even if it is profound. They conduct their debates on the basis of logic and evidence, the canons of which are largely understandable to all of them, and largely, if imperfectly, shared. The question of whether a particular scientific claim is true is both meaningful and important for all members of a scientific community, and needs to be answered as scientists are looking forward to reestablishing the detailed work characteristic of normal science.

A historian, in contrast, looks back to a scientific lexicon that long fell out of use; the beliefs, methods, and practices associated with it are alien to the scientific community active in the historian's own time. The question of *truth* of past beliefs does not arise. A statement made in the new lexicon is a different statement than a statement made in the old lexicon. A statement that can be fully and precisely translated can, of course, be evaluated as true or false, but most interesting scientific statements of past science elude such translation. What they say is ineffable in the later lexicons. So, since past scientific beliefs cannot be simply restated in a modern vocabulary, they cannot be simply evaluated as true or false, either. Rather, the historian's task is to recover past lexicons, beliefs, and practices in order to understand them as reasonable and plausible in their own context, and in order to explain why statements that seem plainly false or even nonsensical to a modern reader justifiably had the status of tautologies in a past science.

Chapter 8: Theory Choice and the Nature of Progress

In an account of scientific knowledge, individuals and groups must be treated differently, an important point that *Structure* sometimes ignored. Scientific development leaves far more room for methodological variation to individuals than is traditionally permitted. The variations that are not permitted are only about those aspects of science that are constitutive of individuals' membership in a scientific community. Group members need a shared language to communicate, which means that they need to share not only reference but also the meaning of words. This takes the form of a shared lexical structure. This structure could be implemented in different ways, however. Scientists do not need to share all beliefs, but they do need to share a structured lexicon that *makes sense* of all of them.

The motor that drives conceptual changes in the natural sciences differs from the motors driving similar changes in practices of other sorts. The community of natural scientists is highly cohesive and sharply demarcated from other groups by training, specialized lexicons, established practices, and shared work.

The rationality of conceptual and belief change in science is not threatened by incommensurability or by methodological relativism, for it is compatible with both. Rationality is always assessed with respect to a given body of evidence, evaluated by available methods. Insofar as the shared basis changes, and insofar as the change includes lexical shifts, rational belief and choice are relative to a particular scientific community in a particular larger historical setting.

The goal of evaluations of scientific beliefs, the appropriateness of describing such evaluations as rational, and the sort of progress such evaluations produce over time, all demand that we acknowledge that the logic of truth claims is essential to the work of scientific communities. We should, however, reject the idea that convergence on truth is the goal of science. While the logic of truth claims is clearly prerequisite to scientific development, this development is driven from behind, away from the inherited lexicon, beliefs, and problems. There is no place for the view that scientific beliefs grow closer and closer to the nature of the real world.

Chapter 9: What's in a Real World?

This chapter was to be primarily concerned with two questions: What gives truth its constitutive role in science, if not correspondence to the real? What could a real world be?[d]

Different scientific communities work in different worlds. A changeable world is required to explain why old theories work. A lexicon provides the ontology of the world to which language applies, and using that lexicon, words do refer to objects in the world. Natural-kind terms are *transparent* when they function properly: they then give us the world. When they fail, they become opaque, and have to be seen as words only.

The account of different worlds, constituted by incommensurable lexical structures, should cautiously be extend to practices other than science. There is a sense in which we always move from one world to another incommensurable world. We do pass between worlds as we pass from home to office or to classroom. There are no smooth transitions between them, and we do damage by failing to notice the thresholds that we cross. (For example: treating one's children as one's students, or vice versa; or treating a family argument as a judicial proceeding.) On such occasions, we, as it were, commit category mistakes. When we transition between the worlds smoothly, when we are correctly handling incommensurable lexicons and situations, we are, in our actual lives, bilingual in a sense.

Epilogue

The developmental approach rightly insists on the centrality of hermeneutic, ethnographic history of science for the philosophy of science. Understanding past science and its development requires overcoming the difficulties posed by incommensurability by recovering the structured lexicons of earlier scientific practices. However, anachronistic or Whig history of science should not be abandoned. Its goal is to explain the success of present-day scientific theories, and so it produces anachronistic narratives in which past science appears as constituted by a series of rationally warranted conclusions and choices, leading to our present scientific theories. Past scientific disagreements are then inevitably seen as conflicts between rational precursors and their irrational opponents. Such narratives are necessary for the formation of present scientific identity. Although the two types of history are incompatible, both are needed, because they serve different functions. Hermeneutic historical narratives allow us to understand the past, while Whig narratives allow us to see the past as *our* past, and to use its lessons in the present.

Appendix

The goal of the appendix is a comparison between the claims and the ways of arguing for them in *Structure* and in *Plurality*. There is a shared core of ideas, there are continuities and developments, but there are also discontinuities and revisions. The main common aspect of the two works is that they both focus on scientific communities, constituted by a shared practice, lexicon, and culture. Philosophical problems concerning meaning, conceptual change, understanding, scientific knowledge, and progress take a different form and invite different answers when raised in terms of groups that tolerate some, but prohibit other, disagreements, than when they are raised in terms of an individual, idealized rational agent.

THE PLURALITY
OF WORLDS

An Evolutionary Theory of
Scientific Development

Thomas S. Kuhn

For Jehane
without whom very little!

ACKNOWLEDGMENTS

Ned Block, Sylvain Bromberger, Susan Carey, Dick [Richard] Cartwright, Josh Cohen, James Conant, Caroline Farrow, Michael Hardimon, Gary Hatfield, Richard Heck, James Higginbotham, Paul Horwich, Paul Hoyningen[-Huene], Philip Kitcher, Jehane Kuhn, Eric Lormand, Richard Rorty, Quentin Skinner, Liz [Elizabeth] Spelke, Noel Swerdlow, and the Group at the National Endowment for the Humanities Institute at Santa Cruz.[a]

PART I

THE PROBLEM

CHAPTER 1

Scientific Knowledge as Historical Product

During the past thirty years, philosophers of science, in increasing numbers, have adopted a new perspective toward their subject matter. Influenced in varying proportions by deep difficulties within the previously received tradition, by the study of examples from history of science, and by the later philosophy of Ludwig Wittgenstein, they have more and more focused on what scientists ordinarily do. Rather than studying science as a timeless body of knowledge, they have shifted their concern to the dynamic process by which that knowledge is generated and changed.[1] They have, that is, viewed science as a practice, one among many.

The new perspective has some striking advantages. Its account of science resembles the actual activity of scientists far more closely than that of its predecessor. In addition, it resolves by dissolution two central difficulties that have for two centuries increasingly clearly confronted the tradition it aims to displace. Traditionally, the concept of scientific objectivity has rested upon the twin assumptions that the truth of individual

---

1. These generalizations and those which immediately follow should, properly, be restricted to English-speaking philosophy of science. The story of Continental philosophy of science, especially in its main French and German versions, is different though perhaps converging. For a perceptive sketch of the long-standing cleavage between these two traditions and of the current possibility of rapprochement, see Gary Gutting, "Continental Philosophy and the History of Science," in *Companion to the History of Modern Science*, ed. R. C. Olby et al. (London: Routledge, 1990), 127–47.

candidates for belief could be tested one by one and that those tests could
be conducted on belief-independent evidence. Despite much effort, neither assumption has yet been justified; discontent with that state of affairs
played a prominent role in motivating the new movement; eliminating the
need for them has been a particular, though not always recognized, contribution of the new movement. To many observers, however, these gains
come at too high a price, for they make it unclear what objectivity can be.
To many of its critics, the new movement has deprived science of its cognitive authority. Its portrayal of the way [that] scientists reach conclusions
about the truth of observations, laws, and theories is said to make those
conclusions subjective or irrational, deeply dependent on time, culture,
and special interests. Again and again the new movement has been indicted
for relativist views of this sort, and a few of its adherents actually do hold
them, a circumstance I had best acknowledge before suggesting the direction my counterargument will take.[2]

   In my view, the charge of relativism is directed to a genuine problem
which it seriously misidentifies. The rationality of a belief has traditionally been held to be relative, at least, to the evidence on which it is based,
thus to time, place, and culture. Many of the tradition's proponents have
gone further, acknowledging coherence with other established beliefs as
itself a rational standard for belief evaluation.[3] It is by no means clear that
the new movement in philosophy of science, some enthusiasts excepted, is
subjectivist or relativistic in a sense that its predecessor was not. The new
movement does, however, challenge the tradition in another and, I think,
far deeper way. What is at issue is not the criteria which govern rational
evaluation, but its goal. That goal has traditionally been taken to be the
discovery of objective truths about the real world. This book is an extended
argument that this way of understanding what scientists do is incoherent,

   2. For an early example of the response of the new movement's opponents, see Israel Scheffler, *Science and Subjectivity*, 2nd. ed. (Indianapolis: Hackett, 1982). For an introduction to the
more sympathetic response to relativism and its kin, see the essays assembled in Ernan McMullin, ed., *Construction and Constraint: The Shaping of Scientific Rationality* (Notre Dame: University of Notre Dame Press, 1988). Much fuller discussion is included in Andrew Pickering, ed.,
Science as Practice and Culture (Chicago: University of Chicago Press, 1992).

   3. See, for example, Carl G. Hempel, *Philosophy of Natural Science* (Englewood Cliffs, NJ:
Prentice Hall, 1966), 38–40, 45–46.

and that recognizing the nature of its incoherence opens the way to a reassertion of the cognitive authority of the sciences.

The path ahead is long and marked by apparent detours. A sketch map before setting out will be helpful. This chapter aims to give body to the claims just made about the central difference between the new movement and the main tradition in philosophy of science. For that purpose it first examines the contrasting ways in which the two sorts of philosophy of science use historical examples, a contrast which suggests the new movement's source and provides essential clues to the basic novelties of its approach. The second chapter illustrates an unexpected difficulty encountered in the attempt to provide examples suitable to the new approach, and the third seeks a preliminary way to understand that difficulty's nature. For that purpose it resuscitates the concept of incommensurability, a concept especially emphasized by a few of the new movement's earliest practitioners, but often ignored in more recent years.[4] Though it has regularly been seen as magnifying the threat of relativism, challenging the objectivity of science, and barring the way to scientific progress, incommensurability proves to be the Ariadne's-thread prerequisite to the understanding of the cognitive authority that science can properly claim.

These first three chapters form part I of this book. From it emerges the claim that members of communities must share something I shall be calling a *structured kind set*, of which the roots are innate, but which is largely acquired through postnatal training. The structure of its kind set encodes the ontology of a community: the sorts of objects, behaviors, and situations which are exhibited in its world. Incommensurability then becomes a relation between the structures of kind sets, and it greatly constrains the extent to which the kind set of one community can be enriched by

---

4. Paul Feyerabend and I independently resorted to the term *incommensurability* in 1962, but the phenomena which motivated us were closely related to those already discussed by N. R. Hanson in 1958. All of us spoke of meaning changes that occurred with the introduction of new theories. See Norwood Russell Hanson, *Patterns of Discovery* (Cambridge: Cambridge University Press, 1958); P. K. Feyerabend, "Explanation, Reduction, and Empiricism," *Minnesota Studies in Philosophy of Science* 3 (1962): 28–97; and Thomas S. Kuhn, *The Structure of Scientific Revolutions*, 2nd ed. (1962; Chicago: University of Chicago Press, 1970). The book will henceforth be referred to as *Structure*. Unless otherwise specified, citations will be to the second edition.

borrowing concepts or their names from the incommensurable kind set of another. Local pockets of incommensurability regularly characterize the relation between the kind set of an older scientific community and that of its successors as well as that between the kind sets of the various scientific communities of a given period. Where knowledge claims made by members of one of these communities make use of concepts and terms within one of these pockets, that claim cannot without residue be translated using the kind set of the other. Understanding those claims requires learning the incommensurable parts of the other kind set and setting them in the place of the corresponding parts of one's own. What results from it is not enrichment of a kind set, but a sort of bilingualism.

When I first began to have thoughts of this sort half a century ago, I was blissfully unaware that no existing theory about concepts and their meanings was compatible with them, a fact from which followed much confusion, my own and others'. But such a theory is for many reasons badly needed, and the three chapters in part II of this book suggest what it might look like. Chapter 4 discusses the biological roots of the concepts of kind and of object, as these have been embedded by evolution in the neural structure of very young animals, including human infants. Chapter 5 considers some relationships between these root structures and human language: in particular, it presents a theory of the meaning of the names of what I shall be calling *taxonomic kinds*, the kinds overwhelmingly most prevalent in everyday life. And chapter 6 turns to what I shall call *singleton kinds*, kinds like mass or force that play a large and steadily increasing role in the development of the sciences and that are ordinarily introduced together with one or more of the universal generalizations ordinarily called laws of nature.

Part III returns to the themes of part I, for which part II attempted a foundation. Chapter 7 examines the process by which newly structured kind sets arise from older ones, leaning heavily on the biological concept of speciation in doing so. Chapter 8 asks what differentiates the motor that drives these changes in science from the motors driving similar changes in practices of other sorts. It thus suggests answers to such questions as [those concerning] the goal of evaluations of scientific belief claims, the appropriateness of describing such evaluations as rational, and the sort of progress such evaluations produce over time. The suggested answers to these questions leave no place for the view that scientific beliefs grow closer and closer to the nature of the real world, yet the logic of truth claims is clearly prerequisite to scientific development. Chapter 9 asks what, if not correspondence to the real, gives it that constitutive role, and in answering asks also what a real world could be. Finally, the book concludes with a brief epilogue discussing the functions of the distinction between what the next section describes as two sorts of history. The time to raise the curtain has come.

I

The family of approaches developed by the practitioners of the new movement are often referred to as "historical methodologies,"[5] and there are good reasons for the title. Historical examples of actual scientific work occupy vastly more space in their work than in that of the group they criticize. These examples provide their evidence for having recovered what I previously called "the actual activity of scientists." As a result, the new movement is often seen, both by proponents and opponents, as empirically based in a way that traditional philosophy of science was not.[6] Its critics even suggest that for properly philosophical concerns the new breed of practitioners have substituted mere description.

There are, however, two oddities about the title "historical methodologies." None of those who began the new movement were drawn to it from history or history of science. Most of them were instead trained in philosophy, and all were drawn to history by prior discontent with the prevailing philosophy of science.[7] A second oddity is more consequential. Recourse

5. For the name, and reasons for its use, see Larry Laudan, "Historical Methodologies: An Overview and Manifesto," in *Current Research in Philosophy of Science*, ed. Peter D. Asquith and Henry E. Kyburg Jr. (East Lansing, MI: Philosophy of Science Association, 1979), 40–54.

6. For a very long time I took that position myself. My central contribution to the new movement opened by disparaging the long-current image of science "drawn, even by scientists themselves, mainly from . . . the textbooks from which each new generation of scientists learns to practice its trade." My work, I suggested, would instead display the "quite different concept of science that can emerge from the historical record of the research activity itself." See *Structure*, 1.

7. I think particularly of Paul Feyerabend, N. R. Hanson, Mary Hesse, and Stephen Toulmin. Michael Polanyi, who also contributed, was neither a philosopher nor a historian, but a scientist who took his examples from his knowledge of contemporary practice. I too was trained as a scientist, and my concern with philosophy antedated by almost a decade my discovery of

to historical examples has been standard in philosophy of science for many years: what distinguishes the new movement is not so much its recourse to such examples as the form it gives them.[8] Both the old and [the] new forms are historical, and both have essential functions to which the epilogue of this book will return. But the functions served by the two kinds of history are different and fundamentally incompatible, and that difference calls forth correspondingly different accounts of both the nature and the authority of scientific knowledge.[9]

To illustrate these differences and catch a first glimpse of what it places at risk, consider two sorts of account of Torricelli's discovery that nature does not abhor a vacuum. The traditional sort tells the story of the replacement of erroneous belief by sound knowledge. It shows science developing *toward* a predetermined goal, the real state of affairs, still unknown when the events in the story occurred but now described in authoritative textbooks. I give two versions of this sort, the first from the 1958 reprint of W. Stanley Jevons's classic *Principles of Science*, a book first published in 1874:

---

the relevance of history to it. What drew us all to history to remedy philosophical discontents I am uncertain, but some scraps of evidence may be worth recording. The crucial episode for me is described early in the next chapter, and it promptly led me to Alexandre Koyré's *Études galiléennes*, 3 vols. (Paris: Hermann, 1939), a work that influenced me deeply and was almost certainly known to some of the others. Stephen Toulmin, at least, was deeply influenced by the early chapters of Herbert Butterfield's *The Origins of Modern Science* (London: G. Bell, 1949), which were, in turn, much influenced by Koyré. (The opening page of Butterfield's chapter 1 emphasizes that the central changes in the science of the sixteenth and seventeenth centuries were "brought about, not by new observations or additional experiments in the first instance, but by transpositions that were taking place in the minds of the scientists themselves," almost a paraphrase of Koyré's most central theme.) Gutting's "Continental Philosophy and the History of Science" will suggest that Koyré's training in the Continental philosophical tradition especially equipped him for his likely role.

8. For an extended examination of the use of historical examples in the philosophy of science and for the way in which their principles of selectivity and their form are shaped by philosophical position, see Joseph Agassi, *Towards an Historiography of Science* (The Hague: Mouton, 1963).

9. The tension between these two sorts of history is not restricted to history of science. Herbert Butterfield, *The Whig Interpretation of History* (London: G. Bell, 1931), provides an elegant discussion of the general case. But those who, like Butterfield and me, inveigh repeatedly against Whig history have tended to overlook its constitutive role within the historical process.

The followers of Aristotle held that nature abhors a vacuum, and thus accounted for the rise of water in a pump. When Torricelli pointed out the visible fact that water would not rise more than 33 feet in a pump nor mercury more than about 30 inches in a glass tube, they attempted to represent these facts as limiting exceptions, saying that nature abhorred a vacuum to a certain extent. But the Academicians del Cimento completed their discomfiture by showing that if we remove the pressure of the surrounding sea of air, and in proportion as we remove it, nature's feelings of abhorrence decrease and finally disappear altogether. Even Aristotelian doctrine could not withstand such direct contradiction.[10]

The second version of the traditional account particularly repays attention. It is from a widely respected elementary text by C. G. Hempel, one I regularly use in my own teaching. Published in 1966, when the new movement in philosophy of science was already underway, it differs from Jevons's only in its greater accuracy and circumspection.

As was known at Galileo's time, and probably much earlier, a simple suction pump, which draws water from a well by means of a piston that can be raised in the pump barrel, will lift water no higher than about 34 feet above the surface of the well. Galileo was intrigued by this limitation and suggested an explanation for it, which was, however, unsound. After Galileo's death, his pupil Torricelli advanced a new answer. He argued that the earth is surrounded by a sea of air, which, by reason of its weight exerts pressure upon the surface below, and that this pressure upon the surface of the well forces water up the pump barrel when the piston is raised. The maximum length of 34 feet for the water column in the barrel thus reflects simply the total pressure of the atmosphere upon the surface of the well.

It is evidently impossible to determine by direct inspection or observation whether this account is correct, and Torricelli tested it indirectly. He reasoned that if his conjecture were true, then the pressure of the atmosphere should also be capable of supporting a proportionately shorter

10. W. Stanley Jevons, *The Principles of Science: A Treatise on Logic and Scientific Method* (1874; repr., New York: Dover, 1958), 666–67.

column of mercury; indeed, since the specific gravity of mercury is about 14 times that of water, the length of the mercury column should be about 34/14 feet, or slightly less than 2½ feet. He checked this test implication by means of an ingenious simple device, which was, in effect, the mercury barometer. The well of water is replaced by an open vessel containing mercury; the barrel of the suction pump is replaced by a glass tube sealed off at one end. The tube is completely filled with mercury and closed by placing the thumb tightly over the open end. It is then inverted, the open end is submerged in the mercury well, and the thumb is withdrawn; whereupon the mercury column in the tube drops until its length is about 30 inches—just as predicted by Torricelli's hypothesis.[11]

Like Jevons, Hempel concludes by referring to subsequent observations in which reducing pressure on the open mercury surface (in this case, by carrying the barometer up a mountain, the Puy de Dôme) resulted in a reduction of column height.

These versions of the Torricelli story display two closely related characteristics. Each is, in the first place, narrowly focused on a single episode, which it presents as providing empirical evidence for—reason to believe in—a particular hypothesis, a theory or lawlike generalization. In this case, the episode is Torricelli's experiment with the mercury-filled tube, and the hypothesis is the air-pressure explanation of phenomena previously explained by nature's putative abhorrence of a vacuum. What the experiment provided was a test of the hypothesis. If the mercury in the inverted tube had not behaved as it did, the air-pressure hypothesis would have become vastly more difficult to sustain.[12]

Given this focus, only the hypothesis and the observations which test it play a significant role in the story. Other details are window dressing,

---

11. Hempel, *Philosophy of Natural Science*, 9.

12. Both Jevons's and Hempel's versions are intended to illustrate the so-called hypothetico-deductive method, in which the hypothesis is articulated in advance of the experiment it tests. The other main empiricist methodology, the inductive method, reverses the order, drawing the hypothesis from previously acquired evidence. The focus on the relation between some particular experiment and its hypothetical explanation is, however, the same in both, and it is this focus that is presently important.

and the various versions are free to differ about which and how many they select for inclusion. The tests would, for example, have played the same role (supported the theory to the same extent) whether or not Aristotle's theory (Jevons) or Galileo's "unsound" explanation (Hempel) had existed. Equally gratuitous is the source of Torricelli's hypothesis. Both Jevons and Hempel give that role to the observation that suction pumps could raise water only 34 feet; but it might equally have been played by some other observation or have been omitted entirely; its historical role as stimulus has no bearing on the authority the hypothesis gains from the experiment. Nor, finally, does anything depend on the way in which these various observations led into one another. Given the hypothesis together with water pumps, mercury, and glass tubing, the observations could, with equal effectiveness as tests, have been made in any order, at any time, and in any place.

The second and more significant characteristic shared by these versions of the Torricelli story is the assumption which legitimizes their narrow focus. The purpose of tests, taken for granted throughout, is to determine whether or not a given law or theory is true, corresponds to the world as it really is, independently of what scientists may believe about it. That assumption embodies the so-called correspondence theory of truth, which most have found inescapable. No account of scientific knowledge may depart from it without accounting for its apparent inevitability. But, right or wrong, the correspondence theory need not predetermine the structure of a putatively historical narrative, and that is its place in the stories so far examined: there is a real world; the sea-of-air theory is a hypothesis about its nature; the experiment with a mercury-filled tube is a test of the hypothesis's truth. Nothing matters but the relation between the experiment and the real world. The source of the hypothesis, the historical context within which the test was performed, and the prior beliefs of the people who performed and evaluated it are irrelevant to the evidential status of the experiment.[13]

---

13. It has been standard to describe these irrelevant factors as belonging to the "context of discovery." Philosophy of science, it is then said, is concerned only with the "context of justification." [See Hans Reichenbach, *Experience and Prediction: An Analysis of the Foundations and the Structure of Knowledge* (Chicago: University of Chicago Press, 1938).]

Contrast these two versions of the traditional account with the sort of account provided by practitioners of the new, developmental philosophy of science. Their stories are about changes of belief, and many stages are ordinarily required before the textbook belief is established. They show science developing, not *toward* the still unknown state of the real world, but *away* from some historical (and correspondingly contingent) set of beliefs about that world. Lacking a manageable published version of the Torricelli story, I resort to one I have repeatedly used myself when teaching philosophy of science.[14] Though it starts at the same point as Jevons's version, and thus earlier than Hempel's, it requires, by its nature, far more space to recount.

Aristotle and his immediate followers believed in the impossibility of a vacuum. On that basis they explained a variety of known natural phenomena including: the operation of siphons and pumps, the adhesion of polished plates, and the retention of liquid in a tube open at the bottom but plugged by a thumb at the top. Those explanations were still standard at the beginning of the seventeenth century, though doubts were increasingly expressed about whether a vacuum was in principle impossible. By late antiquity, for example, many natural philosophers believed that there could be a vacuum dispersed between the ultimate particles of matter: in their view only an extended vacuum was impossible. And during the Middle Ages it was generally conceded that God could create an entirely empty region of space if he chose, though no natural or human forces could do so. By the sixteenth century it was also known, though not usually to the philosophers who discussed the void, that water pumps generally failed to raise water to heights much greater than thirty feet. No one took the failure to present a problem of principle, however: pump shafts were hollow logs; pistons were rags wrapped on wooden rods; major leakage was unavoidable.

Galileo was the first to posit a relationship between the failure of pumps and the laws prohibiting a vacuum. He was concerned with the strength of natural materials, and he believed—by analogy with the

14. This version is drawn primarily from the brilliant monograph by Cornelis de Waard, *L'expérience barométrique Ses antécédents et ses explications* (Thouars, France: Imprimerie Nouvelle, 1936), 96–115. [I believe that the long quotation that follows in the main text is Kuhn's translation of de Waard's book.—Ed.]

cohesion of flat plates—that at least part of their cohesive strength was due to the vacuum dispersed among their parts. But material bodies could be pulled apart, their cohesive force overcome, and it therefore appeared that nature's abhorrence of a vacuum was limited. The same limit, Galileo suggested, was manifest in the failure of water pumps above thirty feet. That failure was a measure of the weight that a vacuum would support and thus of the limited power of the vacuum.

Galileo's hypothesis, first published in his *Two New Sciences* in 1638, was greeted with much scepticism even among his admirers. Perhaps a vacuum might be made, but not by the operation of normal machinery. A group at Rome decided to test Galileo's idea. To avoid leaks they replaced the wooden pump shaft with a pipe of lead to the top of which they cemented a glass globe. The pipe was first filled with water and then raised to a vertical position with its mouth immersed in a water vat. The water it contained then separated from the top and stood at roughly 34 feet, sustained there by the force of the vacuum. Galileo was vindicated.

The Rome experiment was performed in 1640 and word of it soon reached Galileo's pupil Torricelli. He undertook to repeat it with heavier liquids, reasoning that, if the vacuum were what supported the liquid column, all columns of given cross-section would break at the same weight and thus at lengths inversely proportional to a liquid's density. Only when he announced the outcome of the mercury experiment in 1644 did Torricelli suggest that it was due to the weight of the atmosphere, a conclusion probably suggested by similarities between the work he was then doing and his previous experience with hydrostatics ("We live," he then wrote, "at the bottom of a 'sea of air'"). Without recourse to the considerable extant knowledge of hydrostatics, especially the work of Archimedes, neither Torricelli's work nor its reception would be comprehensible. His contribution is most aptly described as the transposition into the already well-known domain of hydrostatics of a set of phenomena previously conceived in a way to which hydrostatics was irrelevant.

Like its more traditional predecessors, this account can be extended to include the later work of the members of the Accademia del Cimento as well as the experiment at the Puy de Dôme. But the chief difference between it and its predecessors is already apparent. This is not the story of an event, Torricelli's discovery, but of the extended process which led, *simulta-neously*, to both that discovery and its interpretation. It recounts a series of closely linked stages, between which the cognitive distances are small: from the impossibility of a vacuum; to the impossibility of an extended vacuum; to the impossibility of human agency's producing an extended vacuum; to the demonstration that man can overcome nature's abhorrence of a vacuum (the water column's breaking under its own weight); to the mercury version of the water experiment; and, finally, to Torricelli's hypothesis of the sea of air. Each of these stages is historically situated in time and place. Each pre-pares the way for its successor by providing a position from which that next stage can readily be reached and by comparison with which its merits can be evaluated. Given a suitable time from which to begin the story (condi-tions for suitability are discussed in the next chapter), all its later stages play essential roles. If one or another had been different, that difference would ordinarily have been reflected throughout the later stages of the story.

It is these differences in form between the two sorts of examples rather than differences in their factual content that prove philosophically con-sequential. Both display science advancing by the exposure of new hy-potheses to the results of observation, thus testing them either directly or, more often, by putting them to use. In both cases, furthermore, the tests involve the comparison, mediated by observation, of the new hypothesis with something else. But that something else differs profoundly in the two cases. Much of this book will be required before that difference can be ar-ticulated fully: here I can only suggest the direction the argument will take.

In the form the developmental movement gives to examples, the achieved outcome requires a temporally extended series of comparisons. The terms in these comparisons are, furthermore, two bodies of belief, one actual or currently in place, the other a candidate for its replacement. Ob-servational consequences can be deduced from each, and, if one set of con-sequences fits observation better than the other, the difference favors the former's acceptance. In the form given to examples by the tradition, on the other hand, there is only a single comparison and only one of its terms is a body of belief[s], something from which consequences can be deduced. The other term is reality. It is not a set of beliefs, but rather something one has beliefs *about*: its relation to observation is causal rather than deductive. Observations are only clues to it, and any presently given set of clues is compatible with many different conceptions of reality. For the tradition, in short, current hypotheses are to be compared with a still incompletely known reality toward which, it is often said, successive scientific hypotheses move closer and closer. For the developmental movement, on the other hand, they are to be compared with the body of belief[s] they aim to replace. One view sees science as pulled from ahead by the reality which a body of true belief would capture, the other as pushed from behind, away from what was actually believed in the immediate past. Both are, if you will, evolutionary, but the first is teleological, directed toward a preexisting goal, the second Darwinian.

That substitution of "evolution-from-what-we-do-know" for "evolution -toward-what-we-wish-to-know" is the developmental movement's central innovation. Close attention to historical examples brought it into being.[15] But the achievements of the new movement and the challenges it continues to face are, as the preceding discussion has implied, largely independent of the particular descriptive information such examples provide. Instead, they result from the replacement of individual illustrative experimental events by an extended narrative displaying a number of them with their interconnections. Insistence on the need for such narratives constitutes the developmental, historical, or evolutionary perspective, and this perspective in turn is constitutive of the new movement.

II

As previously indicated, the new perspective resolves two traditional problems long confronted by the empiricist tradition, most acutely in its

---

15. The quoted phrases are from the closing pages of the original edition of *Structure* (p. 170 in that edition, p. 171 in its successor), where they accompany a brief discussion of parallels between the difficulties first presented by Darwinian evolution and those currently presented by the developmental approach to science. The first removes God's plan as the directive force in biological evolution; the second does the same for the causal role of the external world in the evolution of scientific ideas. In *Structure*, where the force of the argument depends primarily on multiple historical examples, the evolutionary parallel is introduced almost in passing, to postpone dealing with the problem of scientific truth. In this book, where historical examples play a much-reduced role, the parallel enters at the very start as a characteristic intrinsic to any nonteleological developmental process, and its exploitation plays a large part of the role previously taken by examples.

twentieth-century forms: the need for a neutral observation language and the reduction of Duhemian holism. Difficulties with them have appeared to challenge the objectivity of cognitive evaluations, and both seem, when viewed from the developmental perspective, to vanish in a way so easy and natural as to carry great persuasive force. The first of them is the older and more central, traceable at least to the seventeenth century, when modern science and philosophy were born. Science was to be the paradigm of sound knowledge. As nearly as sound method could ensure, its conclusions were to resemble in their certainty those of mathematics. For that to be the case, the evidence on which they rested had to be objective: independent of prior belief and of individual or cultural *idiosyncrasy*. It was required, that is, either to be innate, a priori, or else to be purely sensory, dependent only on the biological makeup of the sense organs shared by all normally equipped human observers. Evidence able to satisfy the latter requirement must, it was thought, be compounded from pure sensations—a color seen, a warmth felt, a sound heard. These sensations provided the simple ideas of sense from which, in Locke's pioneering version of the viewpoint, the complex ideas of bodies in the physical world were to be constructed.

Discovering neutral atomic sensations from which the molecular sensation of a chair or a billiard ball could be compounded presented great difficulties, and another was added to them in the early years of this century. The conclusions deduced from laws and theories take the form of statements about what should be observed if those laws and theories are true. Statements can, however—it was then realized—be compared only with other statements, not with sensory observations directly. What objective testing therefore seemed to require was not pure sensations alone but also a neutral observation vocabulary able to express them. It was statements cast in the observation vocabulary that must command assent, independent of culture or prior belief, if the objectivity of science was to be preserved.[16] Pursuit of such a vocabulary has been the central task of the

---

16. A far fuller and more nuanced account of the transition from ideas to statements is included in Ian Hacking, *Why Does Language Matter to Philosophy?* (Cambridge: Cambridge University Press, 1975).

tradition through most of this century, and it remains to be achieved. Most philosophers now doubt that it will be.[17]

Within the developmental philosophy of science there is no comparable difficulty. What requires evaluation is not a single set of beliefs considered in isolation but the *comparative* merits of the *actual* positions which coexist at a particular time and place. Nothing then depends on stating the items to be compared in a neutral, belief-free vocabulary. The vocabulary need only be shared, independent, that is, of the beliefs with respect to which a choice must be made. What must be shared, furthermore, is not the whole of the vocabulary that each requires but only the part of those vocabularies in which observations are reported. And even there, overlap need not be complete. At least during the period of evaluation, those conditions are inevitably fulfilled: the two vocabularies coexist, one of them recently developed from the other and still largely coextensive with it. Only later, as the new set of beliefs continues to develop, do the vocabulary differences between it and its predecessor increase to a point that might bar evaluative comparison. But by that time the required evaluation has long been made, and its results are in place. In short, when evaluation is seen as a historically situated choice between a traditional body of knowledge and its largely overlapping potential successor, the need for a neutral observation vocabulary disappears. Even if such a vocabulary existed, its use would alter neither the decision made nor its legitimacy.

That is the first of the traditional problems dissolved by the developmental approach. The second, commonly known as the Quine-Duhem thesis, disappears in much the same way. For present purposes, it is Duhem's form of the thesis that is relevant.[18]

17. Impressed by its insuperable difficulties, Otto Neurath and Karl Popper in the early 1930s abandoned the search for a neutral observation language. In their quasi-evolutionary versions of the tradition, the language actually used by practitioners of a science to report observations is accepted for use in hypothesis evaluation and is allowed to change with time. To this extent, their views anticipate the new movement as do those of W. V. O. Quine. What they all omit, however, is what, in retrospect, appears the key step. The merits of a hypothesis (or, better, of the body of belief[s] which includes it) are still to be evaluated in isolation; comparison with the views which would be displaced by accepting the hypothesis plays no essential role.

18. The name "Quine-Duhem thesis" conflates two sorts of holism which, though related, are by no means the same. Quine's holism, to which I return briefly in chapter 3, is a form

Return to Torricelli's experiment will indicate what is involved. Whatever his original intention, Torricelli's barometer experiment does serve as
a test of the sea-of-air hypothesis. But the results of the experiment are not
deducible from the hypothesis alone: auxiliary hypotheses are required as
well. The whole of hydrostatics must be premised together with a series of
lawlike generalizations about weight and density, about the instruments
suitable to measure them, and about the role in those measurements of the
medium in which measurement takes place. Does wood immersed in water
have the same weight as in air, no weight at all, or negative weight (all views
that have been held)? In addition, characteristics of air and its containers
must be presupposed: that the sea of air has an upper surface; that air has
weight and can be treated in the same way as water; that vessels able to
contain air are impermeable also to the still-subtler fluids used to explain
a variety of other natural phenomena; and so on. If the barometer experiment had failed to give the expected result, one need not have concluded
that the sea-of-air hypothesis was at fault. Doubtless it was the most likely
suspect, but the difficulty could have been due to any one of the auxiliary
hypotheses. Tests are inevitably holistic; there is no apparent way to test
individual hypotheses in isolation.

Though techniques for reducing the effects of Duhemian holism continue to be discussed in traditional philosophy of science, no generally accepted solution has been found, and greater success in the future seems
unlikely. Within the developmental philosophy of science, however, the
problem disappears. If the goal of evaluation is a choice between two
specifiable bodies of knowledge, then what needs to be tested is only the
generalizations and singular statements about whose status the two differ.
That is a very small fraction of the whole: the new hypothesis itself plus
whatever statements in the older body of knowledge require adjustment to
make room for it. All the rest—the statements about which adherents of
both bodies of knowledge agree—may be called upon freely in the process

---

of meaning holism which leads him to urge the elimination of the very concept of meaning.
Duhem is unconcerned with meanings, and his holism, to be described at once, is much more
modest.

which legitimates choice even though any of those statements may later be at risk in some other evaluation.

Hydrostatics, for example, was common ground for participants in the debate about Torricelli's hypothesis, and could therefore be premised. That the form of hydrostatics which participants shared required later recasting, especially with respect to the notion of weight, did not make the debate less rational. Questions about air, on the other hand—whether it had weight; whether it could be treated as a fluid; whether its absence from the space above the mercury indicated that that space was empty—these and other questions of the sort were not settled. The debate over the Torricelli experiment was a debate about them as well, and the conclusion of the debate changed views not only about the void, but about them also. Fortunately, enough was shared to bring the debate to a conclusion. Where a *choice* between bodies of knowledge is at stake, neither Duhemian holism nor the possible falsity of shared beliefs is an obstacle to a rational outcome.

III

These two long-standing difficulties vanish within the evolutionary approach. But other characteristic aspects of the tradition vanish with them, and their absence has, as stressed above, seemed to challenge the traditional cognitive authority of science. In particular, they have led to repeated charges of relativism. But the relativism with which the new perspective is charged is of two very different, though often conflated, sorts. The first, which is a genuine relativism, I shall for the moment call methodological. If it threatens the vaunted objectivity of science, which I doubt, the threat is not a new one, and I shall here deal with it only briefly. The other, which has been called relativism with respect to truth, does present a clear threat, but not one properly described as relativism. What is at stake is the traditional notion of scientific truth itself.

In the absence of an Archimedean platform, a fixed position immune to the vagaries of time and culture, the outcome of the evaluation of a candidate for knowledge must depend on the particular historical situation within which the evaluation is made. A proposal judged acceptable under one set of circumstances may therefore be judged unacceptable under another, and historically such changes in rational evaluation can be seen to occur. Consider, for example, the very different receptions accorded the proposal for a heliocentric universe in Greek antiquity and in early modern Europe. How, it is then asked, can judgments of which the outcome varies with time and place properly be called objective? Are they not better seen as subjective, perhaps even as irrational? Questions of this sort illustrate the problem of methodological relativism.

This sort of relativism is undoubtedly real, and it does present problems. But those problems are not new. The soundness or rationality of scientific conclusions has, for example, always been relativized to the evidence at hand when they were drawn. Relativizing to accepted belief as well— something the developmentalist must do—raises no additional problems with respect to subjectivity or rationality, and has in any case often been accepted by proponents of the tradition. Both the methodological relativism inherent in the developmental approach and the main criteria deployed in making cognitive evaluation[s] are thus of a quite traditional sort. The problems they present do not result from the developmental approach but from the strange way in which philosophers of science (and others) employ the terms *objective* and *subjective, rational* and *irrational.* The concepts which these terms denote need systematic philosophical scrutiny, from which reform is likely to follow. Clues to the nature of that reform will be found in chapter 8, which returns to the problem of evaluation.[19]

The second form of the charge—relativism with respect to truth—is another matter. Some formulations of the developmental perspective, particularly my own, have been taken to imply that not simply the rationality of a scientific conclusion but also its truth is relative to time, place, and culture. On this reading, nature did abhor a vacuum in Greek antiquity but no longer does so; the Earth was then at the center of the universe but has since been displaced by the Sun. Reasons for reading my work in that way

19. See also my "Objectivity, Value Judgment, and Theory Choice," chap. 13 in *The Essential Tension: Selected Studies in Scientific Tradition and Change* (Chicago: University of Chicago Press, 1977), and "Rationality and Theory Choice," *Journal of Philosophy* 80 (1983): 563–70 [reprinted as chap. 9 in *The Road Since Structure: Philosophical Essays, 1970–1993, with an Autobiographical Interview,* ed. James Conant and John Haugeland (Chicago: University of Chicago Press, 2000)].

will emerge forcefully in the next chapter.[20] But, however tempting that reading, it was not the one I intended. What I took to be at risk was not so much the permanence of truth as its nature. I could not, for reasons shortly to appear, reconcile the correspondence theory of truth, on which I had been raised, with what I had come to see as the way in which scientific knowledge developed. But neither could I see a way to account for scientific progress without recourse to it. What resulted was for me a deep predicament.

The nature of that predicament will need much further discussion, but for present purposes it can be quickly sketched. From the developmental perspective, all scientific evaluations are necessarily historically situated and comparative. What is to be evaluated is some actually proposed adjustment in the existing body of belief[s], and those current beliefs unaffected by the proposed adjustment are available as tools in the evaluation. They are accepted by all participants and external to the issues being examined. Though they scarcely resemble the traditional Archimedean platform, no evaluation could take place without them. That some of them may be doubted or discarded later in an altered historical and evidential situation is irrelevant. It would be irrational not to put them to use. That way of understanding rational scientific evaluations is deeply consequential. If the outcome of such evaluations does not depend on the truth or falsity of the shared beliefs which provide their basis, then those evaluations cannot bear upon the truth or falsity of the disputed beliefs which they recommend be accepted. In this situation, the traditional substitution of approximation-to-the-truth or probability-of-being-true for truth in the evaluation of candidates for belief is of no help. Evaluations made from the developmental perspective's moving platform provide no point of entry for truth as correspondence to reality.

20. That reading fitted *Structure* especially closely because of my repeated references, however equivocal, to conceptual reorientations after which "the scientist works in a different world" (121 and elsewhere in the chapter from which the quoted phrase is drawn). Similar phrases will be found in this book, but not until its closing chapter, by which time a basis for them will be in place. On this subject, see Ian Hacking's splendid paper "Working in a New World: The Taxonomic Solution" and also my response to it in "Afterwords." Both appear in Paul Horwich, ed., *World Changes: Thomas Kuhn and the Nature of Science* (Cambridge, MA: MIT Press, 1993), 275–310, 314–19. ["Afterwords" is reprinted as chap. 11 in *Road Since Structure.*]

What is at risk, however, is not truth itself but only truth as correspondence, and there is a long-standing family of philosophical viewpoints which holds that a far weaker notion of scientific truth is sufficient to permit understanding. Under such names as pragmatism or instrumentalism, these viewpoints suggest that scientific laws and theories are to be evaluated not for their correspondence to some hypothetical reality but for their ability to realize some other, more immediate objective: predicting or explaining natural phenomena are two goals that have been suggested; controlling nature is a third. On this viewpoint truth as correspondence continues to govern the evaluation of the descriptive statements current in everyday affairs as well as the observation statements which provide evidence for scientists. But, for the evaluation of theories and of laws whose subject transcends brute observation, one or another looser conception of truth is taken to be sufficient. I shall sketch the two main ones in a moment.

Viewpoints of this sort have played a prominent role in shaping the doctrines to be presented in the book, but their existing formulations will not do. First, those formulations presuppose a solution to the daunting problems of an observation language, discussed above. They require, that is, but do not supply a principled line between observation statements, to which correspondence truth applies, and statements that refer to the hypothetical entities generated by science. Second, even if such a line could be drawn, which few still believe, the looser notions of truth provided for nonobservational statements are too weak to support the needs of the sciences. Truth as correspondence supplied an essential tool that these pragmatic conceptions of truth lack.

The basic difficulty with pragmatic conceptions of truth is their failure to explain the need to choose between incompatible scientific laws or theories. Suppose that the pragmatists were right and that atoms, say, were merely a mental construct useful for purposes of prediction or explanation. Scientists would still need to select them as the proper tool for the conduct of a piece of research or as the accepted basis for an evaluation. It would be irrational to employ in a piece of research or an evaluation both the atomic theory and another theory with incompatible consequences. The development of science depends heavily on the requirement that scientists choose between incompatible laws or theories, and the correspondence theory of truth has provided the basis for that requirement: scientific statements are about matters of fact, and logical principles, like the law of noncontradiction, must therefore apply to them. Pragmatic theories of truth provide no substitute.

One pragmatist theory of truth makes it the limit toward which rational inquiry tends, but that formulation supplies no reason to use laws like noncontradiction or the excluded middle in *current* research and evaluation, and that is where they are needed.[a] A second formulation, which equates truth with warranted assertability, is equally incapable of forcing choice: there are often excellent evidential reasons for asserting each of two incompatible laws or theories.[b] No effective scientific community can endure such inconsistency for long. If it persists, the affected community regularly restores the law of noncontradiction by bifurcation, one group employing one theory, another an incompatible one. But bifurcation is a last resort. Scientific development appears to depend on scientists' resisting the easy fragmentation of their communities. Each of the parties to a dispute over laws or theories must normally devote strenuous effort to finding evidence able to persuade the other. Science proceeds from evidence, and evidence would have no function in the sciences if it could not be used in that way. To date, the best justification for this way of proceeding has been provided by the correspondence theory. By making scientific statements at all levels factual, it has subjected them to such logical laws as noncontradiction.

This is the fundamental form of the dilemma of the developmental philosophy of science. Properly understood, its evolutionary viewpoint is in irreconcilable conflict with the correspondence theory of truth. But it nevertheless needs some cognitive principle which requires scientists to choose between incompatibles, and none is supplied by the available substitutes for the correspondence theory.[c]

September 19, 1994

CHAPTER 2

Breaking into the Past

The last chapter suggested that scientific knowledge must be understood as the product of a developmental process, a product which each generation inherits from its predecessors, adjusts to fit novel circumstances arising in its environment, and then passes on to the next generation, which continues the cycle. For philosophical analysis of such a process, the unit to be examined is a narrative that culminates in the establishment of one or more interrelated items of knowledge. That choice of end point does not by itself determine where the required narrative should begin, and there is, in fact, no single best choice. But those suitable for philosophical analysis must reconcile two conflicting strategic imperatives. On the one hand, where change of belief is the narrative's subject, its beginning and end must be conceptually distant. If the former can from the start be seen to foreshadow the latter, the route between them will be unrevealing. On the other [hand], the temporal distance between the narrative's beginning and end should not be significantly larger than conceptual distance requires. The greater the narrative's time span, the greater the number of issues it must cover, and the additional ones are largely extraneous to the philosopher's purpose.[1]

1. That history is explanatory, I shall simply premise, though it has been elsewhere much debated.

Once a starting point has been chosen, a third imperative emerges, and it is the most important of all. To set the stage for the narrative, one must present the relevant part of the body of belief[s] then current in a way that is at once faithful to the detailed textual evidence and that also renders it "plausible." Only when those beliefs have been made plausible—only after one can understand how and why they were once taken seriously—can one address such questions as the role of reason in the decisions that led to their alteration. And only with that question answered (one answer would be: no role at all) can one inquire about the objectives which such decision-making processes may achieve. The concept of plausibility here invoked will be illustrated repeatedly in the rest of this chapter, but it is likely, at this point, to seem problematic. Ordinarily the plausibility of a belief or set of beliefs is taken to be a function of the evidence for and against it or them. But that cannot, on pain of infinite regress, be what is intended here. The entire first chapter was an extended argument that evidence functions only in the evaluation of change of belief, not of belief itself. If the plausibility of the beliefs held at the start of the narrative were, in fact, a consequence of the evidence that led to their acceptance, then making their plausibility apparent would require examining the beliefs that they, by virtue of that evidence, displaced. That process would, in turn, require the examination of still earlier beliefs, and so on to the beginning of time. The sense of *plausibility* applicable to narrative starting points must be of some other sort.

That "other sort" is well known in everyday life, in the nonbehaviorist social sciences, and in much of Continental philosophy. It is the plausibility that comes with *understanding* of a previously puzzling fragment of text or behavior.[2] Usually its arrival is experienced as the sudden and unexpected illumination sometimes referred to as an "aha!" experience. Usually,

---

2. Understanding and the sort of plausibility it brings with it [do] not make the older beliefs right, nor [do they] eliminate error and inconsistency among them. But what comes to be understood includes the reasons for making the particular mistakes and the reasons they take the form they do. However, for the person seeking this sort of understanding, talk of mistakes and inconsistency must be a last resort. As the examples to follow should suggest, it is easy to dismiss potentially revealing oddities as mistakes or superstitions. Though that diagnosis is not always wrong, the great temptation to make it must be strenuously resisted.

also, the understanding it provides is understanding of the purpose, intention, or meaning of some bit of human behavior: that of the author of a text, that of one or more political actors, or just that of the person or persons to whom one is talking. In the most familiar cases, that understanding ordinarily involves recognizing that one has misunderstood, placed the puzzling behavior in the wrong category. The person who seemed inexplicably angry was only excited. Sometimes, however, the required category is unavailable to the person who seeks understanding, and that is especially likely if the actor or actors whose behavior is puzzling belong to a foreign culture. Finding categories which make sense of that behavior, make it plausible, permit it to be understood, is the characteristic activity of ethnographers. It requires the sort of holistic interpretation of behavior now increasingly referred to as hermeneutic.[3]

To make plausible the beliefs current at the start of a narrative of cognitive development, the historian must become an ethnographer, too, and the ethnographic component of the narrative that results both intensifies the philosophical difficulties described in chapter 1 and provides the tool needed to resolve them. Unfortunately, verbal descriptions are unlikely to communicate what is involved. Ethnographers insist that fieldwork is essential to learning their calling. Working as a historian, I have found no way to teach students the ethnographic component of their prospective craft except [by means of] a fieldwork equivalent: reading old scientific texts with them, pointing out difficulties with their early attempts at interpretation and alerting them to clues to alternate readings. The three examples which follow provide no more than vicarious substitutes for that

3. Among many discussions of hermeneutic interpretation, the one I've found most helpful is Charles Taylor, "Interpretation and the Sciences of Man," initially published in 1971 [*Review of Metaphysics* 25, no. 1 (1971): 3–51], but now conveniently reprinted as the opening chapter in his *Philosophy and the Human Sciences: Philosophical Papers*, vol. 2 (Cambridge: Cambridge University Press, 1985), 15–57. Note, however, that a major role in Taylor's presentation is played by a principled distinction between the human and the natural science[s]. For reservations about that dichotomy see my "The Natural and the Human Sciences," in *The Interpretive Turn*, ed. David R. Hiley, James F. Bohman, and Richard Shusterman (Ithaca, NY: Cornell University Press, 1991), 17–24 [reprinted as chap. 10 in *Road Since Structure*].

experience, but they may introduce the problematic which is the heart of this book.[4]

I

All three examples are from personal experience. The first—the start of my understanding of Aristotelian physics—has special significance for me, because my encounter with Aristotle forty years ago is what first persuaded me that history of science might be relevant to philosophy of science.[5] But

4. These illustrations were originally developed for the opening lecture in a series of three presented in 1980 at the University of Notre Dame. With appropriate changes in frame, I have used them repeatedly in individual lectures since, and one of those lectures has been published as "What Are Scientific Revolutions?," in *The Probabilistic Revolution*, vol. 1, *Ideas in History*, ed. Lorenz Krüger, Lorraine J. Daston, and Michael Heidelberger (Cambridge, MA: MIT Press, 1987), 7–22 [reprinted as chap. 1 in *Road Since Structure*]. In all these earlier presentations, as in *Structure*, I took myself to be illustrating a special sort of episode within the development of science (whence the term *revolution* in my titles), and I conflated the experience of scientists moving forward in time with that of the historian moving backward. I don't now think that [this] viewpoint was altogether wrong, but the conflation surely was. For use here, the examples have therefore been extensively revised and refocused to illustrate what the historian must go through to recapture a plausible past position from which the narrative can begin.

5. At the time I worked entirely from English translations of Aristotle's Greek, something I would not have permitted a prospective professional in history of science to do after I had switched to that field from physics and learned the standards of responsibility of my new profession. Most of what follows dates from that early time, but outside assistance together with the facing-page translations of the Loeb Classical Library have since enabled me to do vast fine-tuning. For the assistance I am especially grateful to my wife [Jehane Kuhn], to D. Z. Andriopoulos, John Murdoch, B. B. Price, Richard Sorabji, Gisela Striker, and Noel Swerdlow, none of whom should be held responsible for the views expressed, but only for improvement in their expression. I am grateful also to Professor Striker for making available to me her copy of "Concepts of Space in Classical and Hellenistic Greek Philosophy," a Dutch dissertation submitted in 1988 by Keimpe Arnoldus Algra to Utrecht University [published as *Concepts of Space in Greek Thought*, by Brill, vol. 65 in the series Philosophia Antiqua, 1994]. On the subjects of *chora*, *topos*, and *kenon*, my conclusions differ significantly from Algra's, but their present much-improved formulation would have been impossible without his authoritative monograph. If it had been available to me before a late stage of my revisions, this discussion of Aristotle might have been further improved.

Notice, however, that the need for discussions of *Greek* words in the footnotes which follow is only very indirectly a consequence of the fact that Aristotle wrote in Greek. The differences I am concerned to illustrate are conceptual rather than linguistic. If Aristotle had written in English (or if he'd been translated into English in his own time), similar discussion of the way apparently familiar terms like *motion*, *matter*, and *place* were used in his texts would have been required. The two examples that follow this one may suggest, however, that if, counterfactually,

there is a less personal reason for giving it pride of place. Unlike many of the more technical examples still to come, this one can be made fully accessible to the general reader. As I begin in the next chapter to analyze the phenomena that this chapter illustrates, I shall therefore need to refer to it again and again. More technical, more obviously scientific examples will also be used, both in this chapter and later, but I shall not rely on them exclusively.

I first read some of Aristotle's physical writings in the summer of 1947, while a graduate student of physics trying to prepare a case study on the development of mechanics for a course in science for nonscientists. It had not been difficult to decide where the case study must culminate. Newton and his successors were technically too difficult for our students: the story I had to tell must close with Galileo. In that case, however, my story must start with Aristotle, the figure whose ideas Galileo repeatedly attempted to replace with his own.

Not surprisingly, I approached Aristotle's texts with Newtonian mechanics clearly in mind. The question I hoped to answer was how much mechanics Aristotle had known, how much he had left for people like Galileo and Newton to discover. Given that formulation, I rapidly discovered that Aristotle had known almost no mechanics at all. Everything was left for his successors, mostly those of the sixteenth and seventeenth centuries. That conclusion was standard, and it might in principle have been correct. But I found it bothersome, because as I was reading him, Aristotle appeared not only ignorant of mechanics, but a dreadfully bad student of the physical world in general. About motion, in particular, his writings seemed to me full of egregious errors, both of logic and of observation.

Yet these conclusions seemed not to match the case. Aristotle, after all, had been the much-admired codifier of ancient logic. For almost two millennia after his death, his work played the same role in logic that Euclid's played in geometry. In addition, Aristotle had often proved an extraordinarily acute naturalistic observer. In biology, especially, his descriptive

---

Aristotle had written in (or been translated into) the vocabulary of an older English, my discussion of conceptual change could have been largely accommodated within my text proper. The role of footnotes would have been greatly reduced.

writings provided models that were central in the sixteenth and seventeenth centuries to the emergence of the modern biological tradition. How could his characteristic talents have deserted him so systematically when he turned to the study of motion and mechanics? Equally, if his talents had so deserted him, why had his writings in physics been taken so seriously for so many centuries after his death? Those questions troubled me. I could easily believe that Aristotle had stumbled, but not that, on turning to physics, he had totally collapsed. Might not the fault be mine rather than Aristotle's, I asked myself? Perhaps his words had not meant to him and his contemporaries quite what they meant to me and mine.

Feeling that way, I continued to puzzle over the text, and my suspicions ultimately proved well-founded. I was sitting at my desk with the text of Aristotle's *Physics* open in front of me and with a four-colored pencil in my hand. Looking up, I gazed abstractedly out the window of my room—the visual image is one I can still recall. Suddenly the fragments in my head sorted themselves out in a new way, and fell into place together. My jaw dropped, for all at once Aristotle seemed a very good physicist indeed, but of a sort I'd never dreamed possible. Now I could see why he had said what he'd said, and why he had been believed. Statements that I had previously taken for egregious mistakes now seemed to me, at worst, near misses within a powerful and generally successful tradition.

Let me now illustrate some of what was involved in my discovery of a new way of reading Aristotelian physics, one that made the texts make sense. A first illustration will be familiar to many. When the term rendered "motion" by translators occurs in Aristotelian texts, it refers to all the changes that can be undergone by a physical body.[6] Change of position, the exclusive subject of mechanics for Galileo and Newton, is only one of a number of subcategories of motion for Aristotle. Others include growth (the transformation of an acorn to an oak), alterations of intensity

---

6. In fact, there are two terms which translators render as "motion" or sometimes as "change": *kinesis* and *metabole*. All examples of *kinesis* are examples of *metabole* as well, but not conversely. Examples of *metabole* include coming-to-be and passing away, and these are not cases of *kinesis* because they lack one end point (nonbeing is not an end point). From now on, I shall use the term *motion* for *kinesis*, reserving *change* for *metabole*. Cf. Aristotle, *Physics*, book V, chaps. 1–2, esp. 225a1–225b9.

(the heating of an iron bar), and a number of more general changes of property (the transition from sickness to health). Aristotle recognizes, of course, that the various subcategories are not alike in *all* respects; but the cluster of features relevant to the recognition and analysis of motion are, for him, the ones applicable to motions of all sorts. In a sense that for him was literal, not metaphorical, all these varieties of change were like each other; they constituted a single natural family. Aristotle is explicit about the features they shared: a cause of motion, a subject of motion, a time interval in which the motion takes place, and two end points in which the motion begins and finishes. Reiterated recourse to the end points of a motion were among the primary oddities which had so struck me in my early (Newtonian) readings of Aristotle's text.

A second aspect of Aristotle's physics—harder to recognize and even more important—is the central role played by the properties of individual physical bodies.[7] I do not mean simply that Aristotle aims to explain the properties of bodies and the ways they change, for other sorts of physics have done that. But Aristotelian physics inverts the ontological hierarchy of matter and properties which has been standard since the middle of the seventeenth century. In Newtonian physics, a body is constituted of particles of matter, and its properties are a consequence of the way those particles are arranged, move, and interact. In Aristotle's physics, on the other hand, the role of matter is secondary. Matter is needed, but primarily as a neutral substrate in which properties inhere and which remains the same as those properties change with time. That substrate must be present in all individual bodies, all substances, but their individuality is accounted for not in terms of characteristics of their matter but in terms of the particular properties—heat, wetness, color, size, and so on—that clothe that matter with form. Change occurs by changing properties, not matter, by removing some properties from some given matter and replacing them with

7. There is no word in Greek corresponding to this use of *property*, but the term nicely catches the categories specified by Aristotle as those with respect to which change can occur. In his *Physics*, III, 1, 200b33, Aristotle lists them as substance, quantity, quality, and place. Change with respect to substance, illustrated in the previous note by coming-to-be or passing away, is included in the class of changes (*metabole*) but excluded from that of motion (*kinesis*).

others. There even appear to be conservation laws which some properties must obey.[8]

A third aspect of Aristotle's physics will, for the moment, complete this first example, bringing it to the point required for comparison with the two that follow. In the absence of external interference, most changes of property are asymmetric, especially in the organic realm, which provides Aristotle's model for natural phenomena. An acorn naturally develops into an oak, not vice versa. A sick man often grows healthy by himself, but an external cause is needed, or believed to be needed, to make him sick. One set of properties, one end point of change, represents a body's natural state, the one it strives to achieve for itself and thereafter to maintain.[9] The motions which bring these properties closer to realization are known as *natural motions*; they are contrasted with the so-called *violent motions* which, due to an external cause, carry a body away from its natural state.

The properties which bodies of some given kind strive to realize are essential to it; they are among the properties which constitute what has come to be called the body's essence and which make the body the kind of body it is.[10] These properties need not be fully realized at all times: only

8. Cf. Aristotle, *Physics*, book I, and esp. *On Generation and Corruption*, book II, chaps. 1–4 [Aristotle, *On Sophistical Refutations. On Coming-to-be and Passing Away. On the Cosmos*, trans. E. S. Forster and D. J. Furley, Loeb Classical Library 400 (Cambridge, MA: Harvard University Press, 1955)]. Note, however, that the standard substitution of the English term *matter* for the Greek *hyle* is in some contexts significantly misleading. There is no better English term: the *hyle* of a body is the stuff of which it is made, hence its matter. But the specification of *hyle*, unlike that of matter, occurs in layers or strata. At the highest level, the bronze of a statue and the wood of a bedstead are Aristotle's favorite examples of *hyle*. But both wood and bronze also have their own *hyle*, some combination or other of the four sublunary Aristotelian elements. And there are passages, especially those involving the transformation of one of the four elements into another, which suggest a still more primitive level of *hyle*, a substrate which underlies all of these. About Aristotle's view of the existence of this substrate there is much controversy, but nothing presently relevant depends on its outcome.

9. It is just because motion is always *of* a body, a substance, that change *with respect to substance* is excluded from the class of motions.

10. The term *essence* derives from medieval Latin translations of Aristotle: there is no equivalent term in Aristotle's Greek. But Aristotle makes repeated use of an adjectival distinction between what, for present purposes, may be called the essential (*kath 'auta* or *to ti esti*) and the accidental (*symbebekos*) properties of a body. These adjectives he attaches to a number of nouns, especially to *eidos* (originally "appearance") and *morphe* (originally "shape"), but also to *soma* ("body"). For contexts in which even this description of the essence/accident distinction

to the fully knowledgeable does an acorn reveal the properties of the oak it will become. But they must all be present potentially at all times, as a seed of the mature form. Nor are its essential properties the only ones a body may exhibit. Both men and oak trees vary in height, shape, and somewhat in coloration. These so-called accidental properties may vary among members of the same kind and are correspondingly useful in telling them apart. But they were of little concern to Aristotle and his successors. Their concern was with essential properties and with the natural motions that revealed many of them. Though accidental properties and violent motions had causes, no special regularities were thought to govern them. Aristotelians did not regard them as subjects for explanation.

Change of position—the sort of change which is the subject of mechanics—displays essence, too. The property which a stone or other heavy body strives to realize is position at the center of the universe; the natural position of fire is at the periphery. That is why stones fall toward the center until blocked by an obstacle and why fire flies upward to the heavens. They are realizing their nature just as the acorn does through its growth.

What underlies that interrelationship is the classification of a body's natural position, its home place or own place, as one of its properties. Place at the center is to a stone what leaf size and shape [are] to the mature oak or what normal body heat is to the healthy man or woman. None of these properties need be realized (the stone may be on a hilltop; an acorn has no leaves; body temperature may be disturbed by illness). But each of these bodies must be characterized by some property of the relevant sort and must strive to realize the one that is natural to it. Making natural place an essential property is consequential. The properties of a falling stone change as it moves: the relation between its initial and final states is like that between the acorn (or sapling) and the oak or between the youth and the adult. For Aristotle, therefore, local motion is a change of state, rather than a state as it is for Newton. Newton's first law of motion, the principle of inertia, then becomes unthinkable, for a state is what endures in the absence

---

goes decisively astray, see the brilliant essay on Aristotle in G. E. M. Anscombe and P. T. Geach, *Three Philosophers: Aristotle, Aquinas, Frege* (Ithaca, NY: Cornell University Press, 1961), 5–65, esp. 5–39.

of external intervention. If motion is not a state, then an enduring motion requires force throughout. Aristotle's assimilation of change of place to other sorts of change will need to be undone and replaced en route to the physics of Galileo and Newton.[11]

In that transformation, changing views about the possibility of a vacuum played an essential role, and I will return to Aristotle's physics at the end of this chapter to see how that could be the case. Before considering that aspect of Aristotle's physics, however, I shall present two other examples, both of them more recent and more clearly scientific than Aristotle's view of motion, and then say something about the characteristics all three share.

II

My second example is the problem of setting the stage for the story of the development of the theory of current electricity. That story begins with Alessandro Volta's discovery of the electric battery in the year 1800. A contemporary historian who attempts to tell it will try initially to understand Volta's papers in terms of the conceptual vocabulary he or she has learned for the battery and current. A few words will recall what this vocabulary is. Figure 1a shows a schematic diagram of a single battery cell: two different metals in a container of liquid. The metals shown are silver and zinc, for these are the materials of the coins with which Volta constructed his first

11. One source of its undoing may already be clear. Except for those changes (like repainting a house) which involve alterations only in purely accidental properties, all the sorts of motion with which Aristotelian science deals carry a body closer to or further from the realization of its essence. Change of place, the single alteration which we call motion, displays this characteristic too, but only for motions up and down, toward or away from the center of the Aristotelian cosmos. Lateral motions, which maintain a body's distance from the center, are neutral with respect to realization or deprivation of essence. Difficulties in assimilating such motions to Aristotelian theory (especially to the natural/violent distinction) are the principal source of Aristotle's well-known difficulties in explaining the flight of an arrow or a javelin, and attempts to deal with them led already in antiquity to the invention of so-called impetus theories, which are the background for Galileo's discussions of motion on an inclined plane and of what is often described as his theory of circular inertia. For ancient impetus theories and their sources, see Richard Sorabji, *Matter, Space, and Motion: Theories in Antiquity and Their Sequel* (Ithaca, NY: Cornell University Press, 1988).

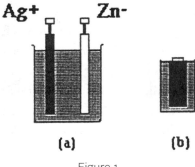

(a)   (b)

Figure 1

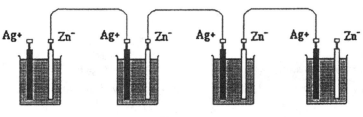

Figure 2

batteries. The zinc then provides the negative terminal of the battery, the silver the positive. If a wire is connected between them, the current which flows through it said to move from plus to minus.[12] To permit comparison, figure 1b shows the familiar flashlight battery. In this battery a central carbon rod replaces the silver of the schematic battery and its outer shell is the zinc. The space between is filled with a granular substance impregnated with the requisite liquid. With these configurations, the strength of individual cells is small. Most applications require more, and it is provided by connecting a number of cells in a chain, the positive terminal of one being attached to the negative of the next, as in figure 2.

12. The choice of direction in which the electric current is represented as flowing is conventional. The actual direction of its physical flow depends on the charge of the particles that carry it; negative particles moving in one direction have the same effects as positive particles moving in the other. Electrically, both constitute the same current. Since the current in metallic conductors is carried by electrons which move from minus to plus, the plus-to-minus convention which I learned in school and am using here is often now reversed.

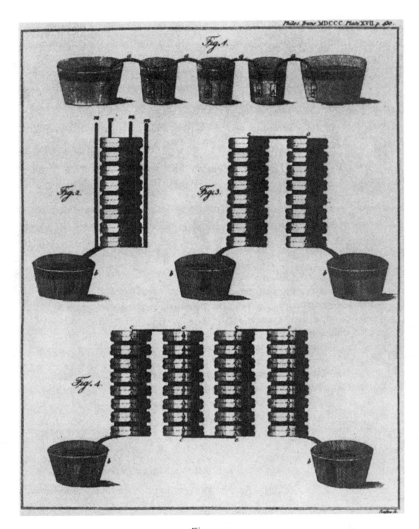

Figure 3

With these concepts in mind, examine the picture, reproduced in figure 3, which accompanied Volta's announcement of his great discovery.[13] It

13. Alessandro Volta, "On the Electricity Excited." On this subject, see Theodore M. Brown, "The Electric Current in Early Nineteenth-Century French Physics," *Historical Studies in the Physical Sciences* 1 (1969): 61–103, and Geoffrey Sutton, "The Politics of Science in Early Napoleonic France: The Case of the Voltaic Pile," *Historical Studies in the Physical Sciences* 11, no. 2 (1981): 329–66. For significant improvements in earlier versions of this example, I am much indebted to conversations with June Z. Fullmer.

was part of a letter, addressed to Sir Joseph Banks, president of the Royal Society, and intended for publication. At first glance it looks familiar, but there are some little-noticed oddities about it. Look, for example, at one of the so-called piles (of coins) in the lower two-thirds of the diagram: reading upward from the bottom right, one sees a piece of zinc, Z, then a piece of silver, A, then a piece of wet blotting paper, then a second piece of zinc, and so on. The cycle zinc, silver, wet blotting paper, is repeated an integral number of times, eight in Volta's original illustration. But *batteries* are not, we think, built that way. The cycle is apparently wrong. If the bottom element in a pile is zinc, it should be followed successively by wet blotting paper and silver rather than by silver and wet blotting paper. Two of the elements constituting the normal cell have been reversed.

These perceived anomalies are not due to mistakes of Volta's, but result rather from our having looked at Volta's diagram through the conceptual spectacles of a later physics. Properly understood, furthermore, they provide an important clue to the recovery of the spectacles he and his contemporaries actually did wear. If one puzzles over the diagram with the aid of the accompanying text, two related misreadings emerge for simultaneous correction. For Volta, the term *battery* refers to the entire pile, not to a subunit composed of a liquid and two metals. (Neither of the isolated cells shown in figure 1 would, for him, have been a battery at all.) Volta's individual subunits, which he refers to as *couples*, do not, furthermore, literally include the liquid at all. Volta's subunit is simply the two pieces of metal in contact. The source of its power is the metallic interface, the bimetallic junction that Volta had previously found to be the seat of an electrical tension, of what we would call voltage. The role of the liquid is simply to connect one unit cell to the next without generating a contact potential which would neutralize the initial effect.

These features are all closely interrelated. Volta's term *battery* is borrowed from artillery, where it refers to a group of cannons fired together or in rapid succession. By his time, it was standard to apply it also to a set of series-connected Leyden jars or condensers, an arrangement that multiplied the tension or the shock that could be gained from an individual jar acting alone. Volta's understanding of such electrostatic devices was very much like our own, and he fitted his new apparatus to the concepts it

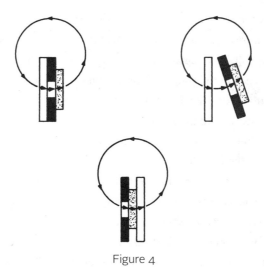

Figure 4

provides. Each bimetallic junction or *couple* is, for him, a self-charging condenser or Leyden jar, and the battery is formed by their linked assembly. For confirmation, look at the top part of Volta's diagram, which illustrates an arrangement he called "the crown of cups." The resemblance to figure 2 is striking, but there is again an oddity. Why do the cups at the two ends of the diagram contain only one piece of metal? What accounts for the apparent incompleteness of the two end cells? The answer is the same as before. For Volta the cups are not cells but simply containers for the liquids that connect the bimetallic horseshoe strips or couples of which his battery is composed. The apparently unoccupied positions in the outermost cups are what we would think of as battery terminals, binding posts. Again, the apparent anomaly in the diagram is of our making.

As in the previous example, the consequences of this electrostatic view of the battery are widespread. For example, as shown in figure 4, the transition from Volta's viewpoint to the modern one reverses the direction of current flow. A modern cell diagram (at the bottom of the figure) can be derived from Volta's (upper left) by a process like turning the latter inside out (upper right). In that process, what was current flow internal to the cell becomes the external current, and vice versa. In the Voltaic diagram the external current flow is from white metal to black, from negative to positive.

In the transition to the modern diagram the direction of flow has been re-versed. Far more important conceptually is the change in the source of the current. For Volta, the metallic interface was the essential element of the cell and necessarily the source of the current it produced. When the cell was turned inside out, the liquid and its two interfaces with the metals pro-vided its essentials, and the source of the current became the chemical ef-fects at these interfaces. During the 1820s and 1830s, when both viewpoints were briefly in the field at once, the first was known as the contact theory of the battery, the second as the chemical theory.

Those are only the most obvious consequences of regarding the battery as an electrostatic device, and some of the others were even more immediately important: for example, the electrostatic viewpoint suppressed the concep-tual role of the external circuit. Only at the moment of discharge are Leyden jars connected to anything except each other. Their discharge path is no more an external circuit than is the path of a streak of lightening, which their dis-charge resembles. As a result, early battery diagrams in the Voltaic tradition do not show an external circuit unless some special effect, like electrolysis or heating a wire, is occurring there, and then very often the battery is not shown. Not until the 1840s do modern cell diagrams begin to appear regularly in books on electricity. When they do, either the external circuit or explicit points for its attachment appear with them. Examples are shown in figure 5.[14]

Finally, the electrostatic view of the battery leads to a concept of electrical resistance very different from the one now standard. There is an electrostatic concept of resistance, or there was in this period. For an insulating material of [a] given cross-section, resistance was measured by the shortest length the material could have without breaking down—without leaking, that is, or ceasing to insulate—when subjected to a given voltage. For a conducting material, resistance was measured by the shortest length the material could

14. These illustrations are from Auguste [Arthur] de La Rive, *Traité d'électricité* [*théorique et appliquée*, vol. 2 (Paris: J.-B. Baillière, 1856)], 600, 656. Structurally similar but schematic diagrams appear in Faraday's experimental researches from the early 1830s [see Michael Far-aday, "Experimental Researches in Electricity," *Philosophical Transactions of the Royal Society of London* 122 (January 1832): 130–31]. My choice of the 1840s as the period when such dia-grams became standard results from a casual survey of electricity texts lying ready to hand. A more systematic study would have had to distinguish between British, French, and German responses to the chemical theory of the battery.

Figure 5

have without melting when connected across a given voltage. It is possible to measure "resistance" conceived in this way, but the results do not conform to Ohm's law. If one is to make measurements that do, one must reconceive the battery and circuit on a more hydrostatic model. Resistance must become like the frictional resistance to the flow of water in pipes. Both the invention and the assimilation of Ohm's law required a noncumulative change of that sort, and that is part of what made his work so difficult for many people to understand and accept.[15] His law has for some time provided a standard example of an important discovery that was initially rejected or ignored.

III

My third example is still more recent and technical than its predecessor. It involves a new interpretation, not yet everywhere accepted, of Max Planck's early work on the so-called black-body problem, primarily as it appears in the famous papers Planck wrote at the at the end of 1899 and the beginning of the year that followed.[16] They contain a derivation of the

15. M. L. Schagrin, "Resistance to Ohm's Law," *American Journal of Physics* 31 (1963): 536–47.

16. For a fuller account, together with supporting material, see my *Black-Body Theory and the Quantum Discontinuity, 1894–1912* (Oxford: Clarendon and Oxford University Press, 1978; paperback ed., Chicago: University of Chicago Press, 1987). A briefer account of the main arguments will be found in my "Revisiting Planck," *Historical Studies in the Physical Sciences* 14, no. 2 (1984): 231–52, reprinted in the paperback edition of the book. For guidance to the sources of the discussion which follows and for the sources of all quotations, see the book, 125–30 and, esp., 196–202.

now familiar law of black-body radiation that Planck had invented a few months before, and modern readers regularly read that derivation as the revolutionary one for which Planck is known. Again, I begin by rehearsing this modern reading that my stage setting aims to replace.

For purposes of his derivation Planck imagined a closed cavity with conducting walls, held at a fixed temperature, and filled with electromagnetic radiation. Within the wall or in the interior of that cavity, he imagined also a very large number of electromagnetic resonators (think of them as tiny electrical tuning forks) each able to absorb and reemit energy at a specified frequency v. Any speck of soot or dust in the cavity enabled these resonators to exchange energy among themselves, increasing the amount at some frequencies and decreasing that at others. By the time Planck took up the problem, it had been known for some years that, in the laboratory apparatus which his model was meant to represent, the radiation would gradually distribute itself in a way that depended only on the temperature of the cavity. Some universal law determined the fraction of the total radiant energy characterized by each frequency. In October 1900 Planck proposed a form for that law which has been accepted ever since. Then, in December of that year and January of the next he provided that law with a derivation.

What has been said to be startling about that derivation (and what was certainly startling about most of its later versions) is that it restricted the energy U_v of any single resonator to integral multiples of hv, where v is the resonator's frequency and h is a universal constant introduced by Planck and later known by his name. That restriction is incompatible with both classical mechanics and classical electromagnetic theory, and its introduction marks the beginning of the end of classical physics. Usually it is embodied in the equation for resonator energy, $U_v = nhv$, and it can, for present purposes, be usefully represented by a diagram like figure 6. The single continuous bar in the top half of that diagram represents the amounts of energy permitted to a single resonator, an energy which may lie anywhere in the interval between 0, no energy at all, and E, the total energy available to the entire collection of resonators. (Neither of those extreme positions is at all likely, but both are mathematically possible, so they must be included in the derivation.) The broken bar in the lower half of the picture represents

Figure 6

the situation said to have been posited in Planck's derivation. Each of the individual blocks represents a single indivisible quantum of energy, ε, of size $h\nu$. Rather than being able to take on any energy between o and E, a resonator of frequency ν must lie at one of the intraquantum breaks in the bar or else, improbably, at one of its two ends.

That understanding of Planck's first papers fits very closely the form which derivations of his law have taken since about 1910. But anomalies arise when these earlier papers are read in this way. They are closely interrelated, and the one which first forced itself on my attention had been previously noticed by the historian of physics Max Jammer. Describing Planck's very first derivation paper, he wrote: "It is also interesting to note that nowhere in this paper, nor in any other of his early writings[,] did Planck bring into prominence the fundamental fact that U [the energy of a resonator] is an integral multiple of $h\nu$."[17] In the event, Jammer's point is even more interesting than he suggests. Not only does Planck fail to emphasize

17. Max Jammer, *The Conceptual History of Quantum Mechanics* (New York: McGraw Hill, 1966), 22. Though made only in passing, this is an especially acute remark. A comparison of my three examples will suggest that the sorts of anomalies which necessitate reinterpretation become more and more difficult to recognize as the date of the text in which they are found becomes more and more recent. As differences in language, typography, and belief are reduced, the reader's expectations force themselves more and more easily onto the text. Though I'd previously read (and forgotten) Jammer's remark, I had to rediscover this difficulty for myself, and I would not have done so if I had not, in pursuit of a different problem, been looking systematically for the point at which Planck first discussed the possibility of generalizing from the quantization of his hypothetical resonators to the quantization of mechanical systems more generally. I had previously both read and taught these early papers without recognizing that Planck did not for some years quantize his resonators at all.

that resonator energy is an integral multiple of $h\nu$, he fails until 1908 even to mention that "fundamental fact," even in his copious correspondence, and by then it had twice been suggested by others. One wonders whether Planck and his contemporaries understood his derivation in the way it has been understood by subsequent physicists and historians.

Two other textual oddities suggest the possibility that they did not. The term *quantum* was, at the turn of the century, regularly applied in German scientific writing to indivisible objects and magnitudes: in particular to the atom, the quantum of matter, and the electronic charge, the quantum of electricity. Planck used it for both of these and also, with some emphasis, for the new quantity h, which he called the *quantum of action* (*action* being a technical term from mechanics). But not until a letter written to H. A. Lorentz in 1909 did he begin regularly to apply the term *quantum* to ε or $h\nu$.[18] In his early papers he did not conceive them as indivisible, as quanta.

The third oddity has not been visible in my presentation, which follows Planck in referring to the hypothetical entities his derivation required as *resonators*. But the word used for them today (and regularly attributed to Planck by historians) is *oscillator*, and there is a difference. Resonators are, in the first instance, acoustic entities. When Planck first introduced them, he indicated that, for the use to which he put them, they might indifferently be conceived as either electric or acoustic. They were thus like tuned strings or pipes, entities which respond gradually to stimulation, the amplitude of their oscillations increasing or decreasing slowly with the size of the stimulus. An oscillator, on the other hand, is simply an object that moves to and fro in a regular cycle. Any resonator is thus an oscillator, but not vice versa (note the use of *oscillation* in the preceding sentences). Shortly after being persuaded that the energy of his resonators must be restricted to an integral number of quanta, Planck began systematically to

18. [The full letter, written on June 16, 1909, is published in A. J. Kox, ed., *The Scientific Correspondence of H. A. Lorentz*, vol. 1 (New York: Springer, 2008), 285–86; a partial reproduction is contained in Kuhn, *Black-Body Theory*, 305n37]. Once in 1905 and again in 1907, Planck did make references to *energy quanta*, but both were in letters addressed to physicists who were themselves using that terminology in discussing his work. [The letters are to Paul Ehrenfest on July 6, 1905, and Wilhelm Wien on March 2, 1907; see Kuhn, *Black-Body Theory*, 132 and 305n44].

banish the term *resonator*. In a letter of January 7, 1910, to H. A. Lorentz, Planck wrote: "Of course, you are entirely right to say that such a resonator no longer deserves its name, and that has moved me to strip it of its title of honor and call it by the more general name 'oscillator.'"[19]

Pursuit of these three anomalies—the absence from Planck's early papers of a restriction on resonator energy together with the conceptual displacement of *resonator* and *energy quantum* by *oscillator* and *energy element*—discloses that all vanish from Planck's work during 1908 and 1909. All of them, furthermore, are first acknowledged in an extended correspondence with H. A. Lorentz, a correspondence initiated by the latter's discussion of the radiation problem at an international congress of mathematicians in Rome in April of the earlier year. It is hard not to conclude that, before 1908, the quantum restriction on resonator energy played no role in Planck's work. Some other way of understanding Planck's early papers is required.

Given that conviction, an alternate understanding of Planck's original black-body theory is readily achieved. Beginning shortly before 1900 his attempts to develop a theory for the distribution of radiant energy were closely modeled on Ludwig Boltzmann's theory of the distribution of thermal energy in a gas. The latter theory was statistical: it required computation of the relative probabilities of different ways of distributing the total available energy over a collection of molecules. For that computation, standard mathematical techniques required that the total energy be conceptually divided into small finite elements of size ε. After the computation was concluded, the continuity characteristic of the physical situation could be restored by allowing ε to become zero.

As indicated in figure 7 (constructed to suggest the inside-out batteries of the previous example), Planck saw himself as proceeding in the same way. For purposes of computation, the energy, E, available to the numerous resonators of frequency v, is mentally subdivided into elements of size ε ($=hv$). Various mathematical distributions of these energy elements over

19. [The original reads: "Freilich sagen Sie mit vollem Recht, dass ein solcher Resonator sich seines Namens nicht mehr würdig zeigt, und dies hat mich bewogen, dem Resonator seinen Ehrennamen abzuerkennen und ihn allgemeiner ‚Oscillator' zu nennen" (Kuhn, *Black-Body Theory*, 305n42; for the full letter, see Kox, *Scientific Correspondence of H. A. Lorentz*, 296).]

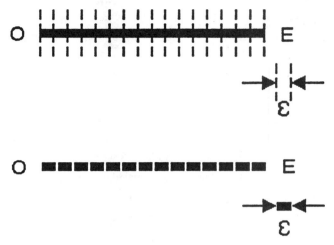

Figure 7

the resonators are considered, but no corresponding physical restriction is applied to them. Planck's resonators, like Boltzmann's molecules, may have any energy at all, lie anywhere between 0 and E on the bar in the top half of the diagram. The equations in Planck's early papers which are sometimes read as restrictions on resonator energy always take a form like $U_N = P\varepsilon$, where P is an integer and U_N is the *total* energy of the N resonators of frequency v. It is that energy, not the energy of individual resonators, which has been divided into small elements *for the sake of computation*. The energy of individual resonators is not restricted.

In the event, of course, Planck's problem is not just like Boltzmann's, and Planck soon encountered an anomaly of his own. The energy element, ε, cannot be permitted to go to zero but must be held equal to hv. Though individual resonators are permitted to move continuously within and across them, the subdivisions of the energy continuum in the upper part of figure 7 are fixed once and for all. Planck later attributed them to what he spoke of as "the physical structure of phase space,"[20] and he hoped to relate that structure to some relationship between the quantum of

20. [Max Planck, "Die physikalische Struktur des Phasenraumes," *Annalen der Physik* 50 (1916): 385–418.]

electricity, e, and the quantum of action, h. But his efforts in that direction were unsuccessful.

When Planck's early papers are interpreted in this way, a number of their better-known anomalies prove also to be eliminable, anomalies which in the past have repeatedly led modern readers to report that Planck did not understand what he was doing or that he was confused. In particular, Planck's failures to follow Boltzmann in computing the most probable distribution of energy over his resonators and his failure to let ε go to zero are readily explained.[21] But these anomalies, though they provide the most fundamental reasons for adopting the interpretation, are too technical for discussion here. Rather than pursue them further, I shall pause to examine two central characteristics shared by the three examples presented to this point. The two are closely related, and they together introduce the central problematic of this book. What follows is a first presentation of issues to which we shall be returning again and again.

IV

The first of these two key characteristics is the easier to describe. The route to all three of these exemplary stage settings begins with the recognition of anomalies in a textual record of past belief. No one can attempt to read such a text without bringing to the task the vocabulary and concepts of his or her own times, and these tools seem for the most part adequate. But in all old texts and many quite recent ones, there are also isolated sentences or passages, equations or diagrams, that don't, on examination, make sense. Typically they are, once noticed, so plainly absurd in present-day terms that it is hard to suppose an intelligent person could have credited them. (Can Volta really have been ignorant of how to build a battery? Can Planck have failed to notice the discontinuities in the energy of his resonators?) Sometimes, instead, their very content is deeply unclear. (What can

---

21. The interpretation sketched above is not logically required for the elimination of these more technical anomalies. But the interpretation makes the path to their elimination vastly easier to find.

Aristotle have meant by the definition "The fulfillment of what exists potentially, insofar as it exists potentially, is motion"?)[22]

Especially in texts from the more recent past (which, for the sciences, can date from as early as the mid-sixteenth century), such passages are extraordinarily easy to overlook. A modern reader knows what the author must have intended and unwittingly adjusts the text to conform; if the anomaly disguised by that adjustment is pointed out, the difficulty is characteristically attributed to the author's "confusion" or to some other lapse of rationality.[23] Most prospective historians of science acquire the ability to recognize and exploit such anomalies only through professional training. Introducing new students to the field, I regularly tell them to remain especially alert to passages that don't make sense and to refrain, except as a last resort, from attributing them to confusion or some equivalent limitation of human reason. Unlike the apparently unproblematic passages through which they are scattered, textual anomalies provide clues, not so much to what the author believed as to the way he or she thought.

The second characteristic of historical stage setting is far harder to describe. It is the one whose presentation ordinarily requires fieldwork or reading texts with students. A first small step toward its delineation is provided by the term *holistic*, used earlier in this chapter to describe the interpretations that set the stage for the beginning of a narrative. More revealing is the description of hermeneutic interpretation, standard in discussions of the subject, as a process which reveals not only the whole that is being reconstituted but also, inseparably, the parts of which the whole is composed. It is this interrelation of wholes and parts, which emerge together and suddenly in interpretation, which lies behind the talk of "aha!" experiences as well as the recurrent recourse to gestalt switches in the early work

22. *Physics*, book III, chap. 1, 201a10–11. The apparent absurdity of this definition was a consequence of the new science of Galileo and Newton, and attempts to make it absurd were sometimes deliberate. In chapter 7 of his *Le monde*, a work written in French, Descartes quotes the definition in Latin, remarking that he gives the definition in that language because he does not know how to interpret it. But Descartes had learned his Aristotle from the Jesuits: his professed ignorance is at best implausible. René Descartes, *Le monde*, in *Œuvres de Descartes*, vol. 11, ed. Charles Adam and Paul Tannery (Paris: Léopold Cerf, 1909), 39.

23. See the discussion of "confusion" in the paper cited in note 14 [Faraday, "Experimental Researches"].

of the pioneers of developmental philosophy of science.[24] What emerges when the interpretive process succeeds—a culmination often described as breaking into the hermeneutic circle—is simultaneously a new set of beliefs and a new set of subjects to have beliefs about. What has been uncovered— excavated, if you will—is a past Archimedean platform or what, in the preceding chapter, I called a *kind set*.

Return briefly to the examples. Aristotle's beliefs were about motion, but the motions which were the subject of those beliefs were changes of all sorts, not just changes of place. That way of categorizing the phenomenal world was not arbitrary but an integral part of a tight-knit view of nature that made properties causally prior to matter: motion then became change of property and realization of essence its driving force. Or consider Volta's battery composed of unit couples (which were not themselves batteries) interconnected by liquid which was external to them. Again, the choice of units was not arbitrary, but an essential component of an electrostatic view of phenomena which came later to be viewed as dynamic. Within the older view, batteries *discharged* through materials that failed, more or less easily, to prevent charge from *leaking*. Within the later view, currents *flowed*, more or less freely, through materials able to *conduct* their charge. Or compare Planck's derivation of the black-body law with its successor. The former distributed resonators (as Boltzmann had distributed molecules) over the elements of size ε into which, for statistical purposes, the energy continuum had been divided. Its successors first replaced resonators with oscillators and mathematical energy elements with physical quanta of energy. Those substitutions, in turn, required a reversal of the direction of distribution (quanta were distributed over oscillators rather than resonators over elements). Taken individually, none of these respects in which Planck's derivation differs from its successors seems to make sense; each is seen to provide evidence of his confusion. But taken together, like the

---

24. Talk of gestalt figures like the duck-rabbit or the old woman–young woman is fundamental to the argument of *Structure*, but they had been introduced for many of the same uses in N. R. Hanson's *Patterns of Discovery*, four years before *Structure* appeared. [Probably the most influential philosophical use of gestalt images predates both Hanson and Kuhn: Ludwig Wittgenstein's *Philosophical Investigations*, first published in 1953, trans. G. E. M. Anscombe (Oxford: Basil Blackwell).—Ed.]

components of Aristotle's or Volta's viewpoint, they provide an extremely close-knit and coherent set of concepts with which to consider the phenomena to which Planck applied them.

These two related characteristics of our examples—the role of anomalies and the close interdigitation of the unfamiliar concepts required to remove them—raise the problem to which this book is, in one way or another, primarily directed. What are we to make of the cognitive status of the doctrines in the older text? In what sense, if any, can we speak of the progress made since it was written? Under the rubric relativism-with-respect-to-truth, a very similar problem appeared in chapter 1. The examples in this chapter, I suggest, have both altered its form and deepened its consequences.

Compare the cognitive evaluation of these out-of-date views in these examples with the cognitive evaluations considered in chapter 1. The latter were performed by the members of a relevant scientific community at some particular time and place. Their purpose was to compare two current but incompatible bodies of belief: one of which had for some years provided the basis for community practice; the other, a new competitor formed by incorporating one or more new beliefs into the first and introducing the adjustments required to make them fit. Most of the body of current belief was common to both, as was most of the current conceptual vocabulary, and that overlap provided the shared Archimedean platform from which community members weighed evidence and conducted evaluation. Without such major overlap, neither activity would have been possible.

The situation presented by the examples is different: two cultures, two Archimedean platforms if you will, are involved, one from the interpreter's time and place, the other from that of the author of the text. Here, too, there must be some overlap between the shared beliefs and concepts that constitute each: in its absence interpretation of one by occupants of the other could not even begin. But these interpretations must bridge a temporal gap. Residents of the earlier platform are dead for those of the later; residents of the second lie in the unimagined future for residents of the first. As a result, the required overlap need support only interpretation, not conversation, much less evaluation. Conversation would serve no function and is in any case impossible. The most that can occur is the vicarious, one-way affair we have been examining: the attempt by contemporaries to let the past speak to them through old texts. The barriers encountered in this attempt are not the same as those encountered in conversations between occupants of a single platform. Interpretation is regularly required to overcome them, and the sense in which it does so proves problematic.

Contemporaries, people who share an Archimedean platform, often disagree about matters of substance, what to believe about some particular object or about the members of the class to which it belongs. They can, however, ordinarily discuss (and sometimes resolve) the relative merits of these positions on the evidence. Misunderstandings are rare, and the categories required to resolve them ordinarily lie close at hand. But the existence of a time interval between the composition of a text and the education of the reader who would assimilate, criticize, or judge it changes the situation, and the extent of the change increases with the length of the interval. Some passages which, at first glance, appear to record beliefs different from the reader's are recognized, on closer inspection, to be anomalous in the sense described at the beginning of this section: no author capable of producing the text in question could plausibly have held beliefs so absurd as those the reader initially attributed to him. Interpretation is required, and its success depends not on the introduction of new arguments but on the acquisition of an interconnected set of concepts not previously in the reader's repertoire. For purposes of understanding the text, these must displace the concepts the reader initially brought to the task. In the event, that change produces more than an understanding of the initially anomalous passages. The reader regularly begins also to recognize the significance of previously overlooked textual details; sometimes he or she can specify in advance how the author will treat a topic in a still unread portion of the text. With respect to the requirements imposed by the text, reader and author have then become very [much] like contemporaries. However vicariously, the reader has gained entry to the author's culture.

The price of that membership is, however, surrender of a previous ethnocentricity. Prior to the understanding produced by interpretation, the reader could find numerous mistakes and falsehoods in the text, errors that could be corrected from the vantage gained since the text was written. After successful interpretation, however, the situation is different. Though the reader may still find mistakes made by the author, they have to be sought. And about many of the passages that previously seemed false, the reader may be at a loss what to say. It is not that those once-false passages have, after interpretation, become true, but that the whole question of right or wrong, true or false, has ceased to seem appropriate.

Look once more at the examples. Aristotle was mistaken about what kept a projectile in motion after it left the hand, a fact noted by his immediate successors. But was he wrong also about what motion is, about the parameters relevant to its specification, or about the marked similarities between, say, a falling stone and a growing oak? Volta was mistaken to suppose that his liquid conductors were unaltered by the battery's discharge. But was he also wrong about the proper construction of a battery or about the direction in which electricity moves during its discharge? And Planck was mistaken that his method of derivation could produce his distribution law. If he had not overlooked an error in one of his approximations, the energy predicted by his distribution law would have been larger by $\frac{1}{2}hv$—the so-called zero-point energy barely detectable by experiments at the time—than the one he is famous for educing. But was he wrong about the properties of resonators or energy elements; and was he wrong or confused about the proper way to employ statistics in his derivation?

The preceding paragraph should, like the discussion of relativism with respect to truth in chapter 1, make problematic not the existence but the nature of scientific progress. Each of the mistakes just pointed out was eliminated in the light of the evidence during the further development of science, and, in the first two cases, awareness of the mistake played a role in its elimination. Surely that is progress, though gained at a cognitive cost which will be central to the discussion in part III. But these mistakes were, in each case, internal to the text being examined. They could have been pointed out to its author, who might, as relevant evidence came to light, have acknowledged and learned from them. Aristotle seems to have felt the difficulties in his projectile theory, for he offers it in several incompatible versions; Volta was from the start forced to defend his conception of the role of the liquid in his batteries; and Planck did acknowledge his mistaken approximation, adding the zero-point energy to his distribution law in what is commonly known as his "second theory."

The anomalies eliminated by interpretation are not, however, primarily due to an older author's mistakes. Rather, they result from the concepts he or she deployed and to the uses those concepts legitimated. To call them mistakes is to reject, not the conclusions in the text, but the parts from which those conclusions are framed, the parts just made visible by hermeneutic interpretation. What could such a rejection mean? Imagine, for example, Aristotle's puzzlement if told that his concept of motion was mistaken: "Motion," he might plausibly say, "simply *is* change of all sort: that is what, in standard use, the term *motion* means. I've tried," he might then continue, "only to clarify that use by spelling out the senses in which coming to be and passing away both are and are not motions. Is it that clarification that you are rejecting?" Or in the case of Volta or Planck, one can imagine the response, "I've simply borrowed the standard term (*battery* or *element*) from another field and demonstrated the advantages to be attained by adapting the corresponding concept to my new application. What can you have in mind in suggesting that I've been wrong or mistaken?" If anything like the standard notion of scientific progress is to be preserved, those questions require firm answers, and I doubt they can be made available.

V

I am, of course, now well ahead of my story. These characteristics of historical stage settings will need to be examined in more detail, and some easy responses to the difficulties they pose will need to be scrutinized and rejected. That is the task of the next chapter. Before turning to it, however, let me present one deeper example of these difficulties, returning to Aristotle for the purpose. Particularly when presented briefly and nontechnically, the texts which provide my second and third examples make it easy to imagine discussing one or another of them with its author as one would with a contemporary. To some readers the differences between us and them seem to be merely about words, resolvable by an easy redefinition of terms. Many of Aristotle's views are, on the other hand, deeply exotic. We are separated from their author by a vastly larger interval of time; much more has changed since his texts were written. Excepting perhaps his view of motion, it is far harder to imagine stepping from the time machine, shaking his hand, and entering into immediate discussion. Just how hard it would be to find a starting point for such discussion will be indicated if I extend the network of interconnected concepts that I truncated arbitrarily above.

As I first told that story, I imputed to Aristotle the doctrine that nature abhorred a vacuum. I suggested, that is, that he believed in a law of nature that prohibited the existence of empty space and which could therefore be used to explain a number of phenomena like the syphon and the cohesion of polished marble plates. There are, however, two sets of reasons why that way of understanding Aristotle's view of the void cannot be right. First, Aristotle had no conception of a natural law, of the sort of generalization philosophers of science call nomic, and there was no place for them in his science. The empirical part of science was, for him, the identification of essences. In this effort, mistakes could be made, recognized, and corrected. But the only scientific generalizations to which that search could give rise were the consequences of essence: with essence established, the force of generalizations was logical.[25] The status of Aristotle's belief that there could be no void was not that of an empirical generalization which might be confirmed or falsified by experiment—Galileo's, Torricelli's, or any others.

The second barrier to understanding Aristotle's views about the void as a rejection of the existence of empty space is that neither Aristotle nor his Greek contemporaries had the conception of space prerequisite to the formulation of such a prohibition. The required conception makes space the container of all physical things: all natural objects are said to be *in* space. But *chora*, the Greek word most often translated [as] "space," is always a local place or region; there must always be something else, one or more other *chorai*, outside, or external to it. More to the point, the term *chora* suggests proper place in a physical or social structure: it is used, for example, in the phrase "to die at one's post"; the *chora* of a native Corinthian

---

25. Cf. the scathing remark about Aristotle's *Posterior Analytics*, book I, in Anscombe and Geach, *Three Philosophers*, 6. Notice that understanding the force of empirical generalizations as logical makes it appropriate to read Aristotle's use of apparently empirical evidence to counter the views of others as refutations by reductio ad absurdum. The point is particularly relevant to his apparently physical arguments against the void, for which see below.

remains Corinth even after he moves to Athens.[26] Both these usages are reminiscent of home place, the place proper to a body. They suggest a sense in which *chora* need not be a term of location at all but is rather one of a body's essential properties, a property it retains potentially even when it is not realized. Aristotle does use the term that way. In his *On the Heavens*, which sketches the structure of the entire cosmos, the central region of Earth and the concentric surrounding shells of water, air, and fire are all *chorai*, the proper places of the elements which, even when displaced, bear their names.[27] The only Greek term that refers to all these regions together is *kosmos*, usually translated [as] "the world" or "the universe," and it—unlike the term *space*—applies to them not as a simple collective but in, and by virtue of, their proper or natural order. Their nature depends on their interrelationships, and those relationships, until well after Aristotle's death, were topological, not metric or even geometric.[28]

The other Greek term sometimes translated [as] "space" but more often [as] "place" is *topos*, a word which overlaps *chora* in that both can apply to a region and can sometimes be used interchangeably. But *chora* differs from *topos* in three presently important respects. First, a *topos* tends to be smaller than (and by implication located within) a *chora*. Second, and more important, *topos*, unlike *chora*, is neutral with respect to naturalness or order. It does locate a body—answers the question, "Where?"—by specifying the place that the body happens to occupy at a particular time. *Topos* is thus one of the body's accidents, a property which, like the color of a boat, can change without altering the body's identity. A body has both a specific *chora* and a specific *topos* at any time, but the two need not

26. I am much indebted to Gisela Striker for this example, though she is doubtful of the use I make of it. Its predecessor is from the Greek *Lexicon* of Liddell and Scott [Henry George Liddell and Robert Scott, *A Greek-English Lexicon*, 9th ed., vol. 2, rev. Sir Henry Stuart Jones and Roderick McKenzie (1940), s.v. χώρα].

27. *On the Heavens*, IV, 312a5, 312b, 3–7.

28. Understanding the relationships between regions as topological rather than metric is characteristic also both of children and of the members of preliterate societies. For the former, see Jean Piaget and Bärbel Inhelder, *The Child's Concept[ion] of Space*, trans. F. J. Langdon and J. L. Lunzer (New York: Norton, 1967). For the latter, see Heinz Werner, *Comparative Psychology of Mental Development*, rev. ed. (Chicago: Follett, 1948), chap. 5. For an analysis of a closely related parallel between children's and Aristotelian thought see my "A Function for Thought Experiments," available as chap. 10 in my *Essential Tension*.

coincide.[29] Finally, unlike *chora*, *topos* is conceived in metric and quasi-geometric terms. In *Categories*, *topos* provides one of Aristotle's examples of continuous quantity, and is discussed in terms of lines, points, and the relation among the parts of a figure.[30] Its contrast with *chora* is underscored by a passage in which Aristotle, arguing that quantities do not have contraries, points out that a plausible but mistaken case to the contrary can be made by thinking of the cosmos, the ordered collection of *chorai*. In the cosmos up and down are contraries, the directions toward the peripheral and toward the central *chorai*, but this sort of contrariety does not apply either to quantities or to *topoi*.[31]

In the *Categories*, Aristotle gives "in the marketplace" or "at the Lyceum" as examples of answers to the question, Where? Presumably, if the question had been asked at the marketplace, appropriate responses would have been "at the butcher's" or "at the wine merchant's." In the *Physics*, preparing

29. In *On the Heavens*, IV, 312b, 3–7, the second of the passages cited in note 26, Aristotle discusses the question [of] whether a body has weight in its own *chora*. The apparently similar question [of] whether a body has weight in its own *topos* would make no sense unless *autos topos* (home place) were explicitly specified. This whole work is especially helpful in discovering the difference between Aristotle's concepts of *topos* and *chora*. The former occurs at 271a5, 26; 273a13; 275b11; 279a12; 287a13, 22; and 309b26—the latter at 287a17, 23; 309b24, 25; 312a5; and 312b3, 7. No brief generalizations quite capture all these entries, but a *topos* is ordinarily concrete, a particular location fully occupied by a particular body: thus the beginning and end points of a motion or (in close conjunction with *kenon*) that which cannot be emptied of body because it would then not be a *topos*. *Chora*, by contrast, may have a body somewhere in it, but it is not associated with any particular body but rather with a region within an ordered whole (or collectively with that whole in its entirety). The passages in which it occurs are usually cosmological and explanatory. Note that the passages at 287a and 309b make use of both terms. Figuring out why both are needed is a useful interpretive exercise.

30. *Categories*, VI, 4b25-5a-15.

31. *Categories*, VI, 6a11–19. Compare *Physics* IV, i, 208b15–26. The latter passage is important and should be read with the nature of the geometry of Aristotle's time clearly in mind. Its subject was not the geometry of space but of abstract figures, shapes. For Euclid a line is a figure bounded by two points, a plane a figure bounded by lines. On this view, for example, the interior of a plane triangle is itself a plane; outside the triangle there need be nothing at all. (Compare Aristotle's finite cosmos: a sphere containing everything there is without even space outside it.) Congruence is determined by "applying" one figure to another, not by translating one through space until it lies on the other. The concept of a Euclidian space can be fitted to Euclid's *Elements*, but it's not to be found there nor is it needed. (Think of Euclidean figures as Platonic Ideas or forms, outside space and time.) Something like it is needed for and implicit in the parametric treatments of the conic sections which are only beginning to emerge in Aristotle's lifetime. Those treatments required an expansion of the notion of geometry.

for the discussion of the void, he exploits the quantitative and geometric aspects of *topos* to carry this process of increasingly precise specification to the limit.[32] *Topos* then becomes "the inner surface of the surrounding body," and Aristotle proceeds quickly to consider the void, which is for him empty *topos*, not empty *chora*, much less empty space.[33]

The Greek word for the void, the noun *kenon*, was coined by the Greek atomists from the standard adjective *kenos*, or empty, regularly applied to containers.[34] When Aristotle in book IV of the *Physics* first considers the concept, he at once points out that, if the void could exist, then "'place' and the 'filled' and 'vacant' would all be one identical entity under varying aspects or conditions of existence," a position that for him is incoherent and that he repeatedly attempts to eliminate by reductio ad absurdum.[35] Some

32. For the limiting process, see *Physics*, IV, ii, 209a33–209b1. For the limiting definitions, see *Physics*, IV, iv, 211a29–31. There is a considerable scholarly literature (for which see Algra, "Concepts of Space," chap. 4) attempting to explain what have been read as inconsistencies between the discussions of *topos* in the *Categories* and in the *Physics*. But I take these "inconsistencies" to be a product of trying to find the Latin Aristotle—from which Europe learned to understand Aristotelian doctrine—in the Greek texts. With respect to spatial questions, however, the Latin and Greek conceptual vocabularies are very different, and the two cannot be accommodated without significant distortion. Both *topos* and *chora* are usually translated [as] *locus* in Latin. The main alternate is rendering *chora* as *spatium*, equivalent to English "span" or "interval," whether of space or of time. Both Latin terms apply to a measured or measurable interval. The seventeenth-century conception of space is clearly foreshadowed. See also note 33 below.

33. Indeed, in the whole of the *Physics*, the term *chora* occurs only four times (208b8, 208b31–35, 209a8, 209b12–15): all are in IV, i and ii, where Aristotle works out the limiting definition of place, and all of them tightly juxtapose *chora* with *topos*, because the differences between them are relevant to the points being made.

34. Their objective in doing so was more nearly logical than physical: the refutation of the Parmenidean thesis that motion was unreal, the world changeless. They too exploited paradox, often speaking of the void as the "what is not," while insisting on its existence and reality. For them, it was a second existent, the other being the atoms, the "what is." So far as one can tell from their fragmentary remains, they did not think of the void as a space-like container, and I think they could not have. *Kenon* appears to have surrounded the atoms without penetrating them, as the water they swim in surrounds fish. This conceptual situation changed markedly with the expansion of the concept of geometry (note [31]) and when Greek philosophical doctrines were translated into Latin (note [32]). *Spatium* is much like the modern term *space*, and the Roman atomist Lucretius is the first person who, to my knowledge, spoke of empty space, *spatium vacuum*.

35. *Physics*, IV, vi, 213a18–20. The translation of this passage in the Oxford version makes the incoherence even more visible than the Loeb version, quoted above: "For those who hold that the void exists regard it as a sort of place or vessel which is supposed to be 'full' when it holds the bulk which it is capable of containing, 'void' when it is deprived of that—as if 'void'

of his arguments are directed to the concept of [the] void itself. He says, for example: "Since the void (if there is any) must be conceived as place in which there might be body but is not, it is clear that, so conceived the void cannot exist at all," and a few pages later he repeats, "In whatever sense the void is identified with 'place,' when 'place' in that sense is shown not to exist the void vanishes with it."[36] For Aristotle, who had defined place as an accidental property of the body that happened to occupy it, asking whether place could exist without some body or other in it was like asking whether a color could exist without being the color of some body or other. His teacher Plato, who took an otherworldly position with respect to natural phenomena, would have answered the latter question Yes!," but the point is perhaps the most central of those on which the more naturalistic Aristotle separates himself from his teacher.

Other reductio arguments of Aristotle's are often read as empirical, but their force is, as previously noted, more nearly logical. If the void could exist, then a region containing matter could be bounded by a region contained nothing at all. But if that were possible, then the Aristotelian universe or cosmos could not be finite. It is just because matter and the cosmos are coextensive that the cosmos must end where matter ends, at the sphere of the stars, beyond which there is nothing at all, neither space nor matter. That finitude, in turn, underlies Aristotle's theory of local motion, both natural and violent. In an infinite universe, any region of space would be as central (or as peripheral) as any other. There would then be no special regions in which stones and other heavy bodies, or fire and other light ones, could fully realize their natural properties. Nor, if a void were possible, could there be violent motions. For the violent motion of a body requires that it be pushed by a neighbor in contact with it; and that neighbor must itself be pushed, and so on in a chain. But a body in a void would have no neighbors

---

and 'full' and 'place' denoted the same thing, though the essence of the three is different" [Aristotle, *Physics*, ed. W. D. Ross, trans. R. P. Hardie and R. K. Gaye, vol. 2 of *The Works of Aristotle* (Oxford: Clarendon Press, 1930), 213a18–20].

36. *Physics*, IV, vii, 214a16–20, and viii, 27–28. Richard Sorabji, to whom I am indebted for the second of these passages, points out that both are directed exclusively [at] the atomists' conception of a void. But in the absence of a concept of space, what other concept was there for him to criticize?

and could neither move [n]or be moved. An unbounded universe would pose similar difficulties for astronomy. The rotating sphere that carries the stars would have to be infinite, and to rotate with infinite speed.[37]

It must now be apparent that, if the story of Torricelli's discovery of the extended vacuum begins with Aristotle, as it should, then it involves a great deal more than an experiment with a mercury-filled tube. A major network of interlocking concepts needed to have been picked apart and rewoven before Torricelli's experiment could even be conceived. One part of that reweaving, which began in antiquity, is worth noting here. Aristotle rejected the void for reasons that were primarily logical: it was, for him, something like a square circle, a contradiction in terms. But the integrity of Aristotle's physical position, his description of the cosmos and its operation, did not require a position about the void quite so far-reaching. Accepting Aristotle's physical views—his description of the cosmos and his account of change within it—did not require that the void be prohibited on logical grounds. Its prohibition could instead be physical, in which case what had to be barred was only the extended void, a place which, like the top of Torricelli's barometer, could contain body, but in fact did not. The extended void carried with it the infinitude of the universe, which could not be reconciled with Aristotle's position. But allowing empty space within the pores of material bodies—a condition which came to be known as the *dispersed* or *interstitial vacuum*—did not raise physical difficulties, and it was widely adopted by Aristotle's ancient and medieval successors.[38] Galileo was among those who exploited it, and his doing so was, as we have seen, vital to the work of Torricelli. Given the scope of Aristotelian physics and the tight interconnections between its parts, the almost simultaneous emergence of these constitutive contributions to a new cosmology and physics is anything but coincidental.

With this background in mind, return momentarily to the problems of truth and progress for whose sake it was introduced. Aristotle's *void* was not our *void* any more than his *motion* was our *motion* or his *matter* our *matter*, and all these concepts were, for him, closely interrelated. Under those

---

37. For this and closely related arguments, see *Physics* IV, v and viii, esp. 214b27–215a24.

38. For the dispersed void after Aristotle, see Sorabji, *Matter, Space, and Motion*.

circumstances, what are we to say about his statement that there could be no void? Was it right or wrong, true or false, correct or mistaken? To me, none of these alternatives seems usable. We need another way to locate ourselves with respect to our past and our past with respect to us.

September 19, 1994

CHAPTER 3

Taxonomy and Incommensurability

The three examples in the last chapter were designed to display the often-dramatic impact on the historian who undergoes the experience of re-capturing past thought by deciphering its textual remains. Though the reinterpretation required is always local—a characteristic sort of anomaly giving rise to a new understanding of a small number of interrelated terms or concepts—its outcome casts a new light on an integrated system of beliefs. Almost all my work as a historian has been inaugurated by experiences of this sort, and for some years they did great violence to the comfortable cognitive ethnocentricity I had acquired from training as a physicist and from reading in philosophy of science. In particular, they made deeply problematic the belief—constitutive for most scientists and many philosophers—that facts were facts, under whatever rubric or in whatever vocabulary they were described. I did not for a moment conclude that Aristotle's physics or Volta's view of the battery was right, nor that either was as good as our own. But neither could I continue to feel that they were simply wrong, mistaken, false. The result, for me, was a compulsion to find a general way of characterizing these experiences and to explore its significance for the nature of knowledge, especially scientific knowledge.

To the problem of characterizing the examples like those in the last chapter I provided a first, but far too easy, answer in *Structure*, the first extended product of my compulsion. Some of the terms around which the historian's reinterpretation turns have, I there suggested, changed their meaning since the texts under discussion were written, and the historian must rediscover what their meanings then were. That answer was not wrong, but it quickly proved both uninformative and in a central respect misleading. Meaning change of some sort is always with us, both over time and, arguably, from individual to individual as well. What each of my examples specifically illustrates is not meaning change in general but a locally holistic transformation of meanings: the simultaneous alteration, that is, of the meanings of a set of interrelated terms no one of which could—if textual coherence were to be preserved—have changed in the way it did without changing the others as well. Presently recognized theories of meaning do not much illuminate changes of this sort, for most treat meaning as attaching to terms one at a time. In this situation my early talk of meaning change was in practice largely empty.

In retrospect I realize also that talk of meaning change was misleading. Though changes in word meaning are central to the examples, only a limited class of terms is involved. I shall be calling them *kind terms* or, sometimes, *taxonomic terms*, for they name the kinds of things, situations, and properties which occur in the world as we know it. Restricting concern to them, as I shall from this point on, makes the goal narrower than a general theory of meaning. In that sense, talk of meaning change claimed too much. But in another sense, it claimed too little. The ability to discriminate kinds and to deploy a selective repertoire of behavioral responses to them [is] not restricted to linguistically endowed populations. Animal communities display them too. Their members do not have kind terms but they may share a structured kind set. I shall then speak of them as having kind concepts, for concepts need not have names. In linguistically endowed populations, of course, all or most of them do, and I shall therefore continue to speak repeatedly of words and meanings in what follows. But what the explication of examples like those given in the last chapter turns out to require is a theory of the way in which the members, both of human and of animal societies, cut up their worlds, a categorization process without which those worlds would not be worlds at all. That process, whatever it may be, surely involves something like meanings. But its roots are prelinguistic, and the meanings it invokes are not, in the first instance, the meanings of words.

This chapter opens with a sketch of the problems faced by traditional accounts of this categorization process and continues with some early steps toward their resolution. To avoid needless circumlocution, I shall treat the problem as concerned entirely with category names and their meanings. Only here and there will it be necessary to remind readers that kind concepts need not all have names.

I

Look first at some inadequacies of currently received views of the meaning of kind terms. Meaning itself has two standard aspects. The first, usually called the *extension* or *denotation* of a term, consists of its referents, the set of things, situations, etc. to which the term refers. The extension of *motion* is all the motions in the world: present, past, or future; the extension of *planet* is Mercury, Venus, Mars, Jupiter, Saturn, Neptune, Uranus, Pluto, and any other bodies that may be found revolving around one or another star.[a] The extension of *chemist* is all the chemists there have ever been or will be; and so on. The second aspect, for which the term *meaning* is usually reserved, is not the things referred to but the characteristics, properties, or features by virtue of which the term is applied to them. The term *triangle*, for example, means "plane figure bounded by three straight lines." This aspect of meaning is usually labeled *sense* or *intension*.

In the examples of the previous chapter, extension and intension change together. The passage from acorn to oak was a motion for Aristotle, and one of the properties relevant to placing it in the class of motions was its possession of two end points. Neither the former nor the latter is the case for us. The units of Volta's batteries are not batteries for us, nor did Volta pick out his unit batteries in the ways we do. Planck's energy element was not a quantum for him or his contemporaries, and the change of theory which made it one also altered the intension of the term *quantum*. Before Planck's work, constancy of size was a criterion for being a quantum; but the size of the quanta he introduced was variable, changing as energy passed from resonators at one frequency to resonators at another.

The intension or meaning of a term has traditionally been taken to be given by a set of features shared by all its referents and, collectively, by them alone. The notion has a plausible basis. It fits the meanings of mathematical terms like *triangle* or logical terms like *implication*, and it fits well enough the meaning of some terms from everyday language. The definition of *bachelor* as unmarried man provides a frequently invoked example. Though it does not capture all the uses listed in a dictionary, it does catch what, for most users, is overwhelmingly the main one. At least as an idealization the equation of meaning with a list of features necessary and sufficient for identification has seemed the natural one. It has not, however, been immune from criticism, and in the last fifty years it has been increasingly discarded.

One reason for skepticism about this traditional view of meaning is the difficulty in matching it to the way people learn and use language. Few of the "definitions" provided by standard dictionaries are sufficiently precise to be applied unequivocally to any and all imaginable presentations. Besides, none are self-contained: all lead back to other nontrivial dictionary entries, and sometimes back from these to the entry from which the search began. As to the use of terms, introspection suggests what much psychological investigation confirms.[1] People unproblematically apply everyday terms like *cat, dog,* or *bird* without being able to supply anything like a list of the defining features of the corresponding creatures. Systematic search for the features that people actually use in making these and other identifications indicates that, though individuals ordinarily make the same identifications, the features they use in making them vary both with circumstance and from one individual to the next. Experts in the classificatory sciences have exactly the same difficulties in finding necessary and sufficient conditions for membership in their much-refined taxonomic categories. In their efforts to understand the species concept, all have renounced the search for features that provide necessary and sufficient conditions for class membership, and some have abandoned talk of features altogether.[2]

---

1. Edward E. Smith and Douglas L. Medin, *Categories and Concepts* (Cambridge, MA: Harvard University Press, 1981) supplies a convenient entry point to much of the relevant psychological literature. Chapter 3, "The Classical View," is especially relevant. J. A. Fodor, M. F. Garrett, E. C. T. Walker, and C. H. Parkes, "Against Definitions," *Cognition* 8, no. 3 (1980): 263–367, is also helpful.

2. A very useful introduction to these issues is provided by Ernst Mayr, "Biological Classification: Toward a Synthesis of Opposing Methodologies," *Science* 214 (1981): 510–16. Other viewpoints, together with far more detail, [are] provided by the essays in Marc Ereshefsky, ed., *The Units of Evolution: Essays on the Nature of Species* (Cambridge, MA: MIT Press, 1992). David Hull, *Science as Process: An Evolutionary Account of the Social and Conceptual Development of Science* (Chicago: University of Chicago Press, 1988), describes both these issues and the

A closely related difficulty with the classical conception of meaning is presently even more central, for it involves points of principle and will, as we continue, impinge on the concept of scientific law. Suppose that kind terms were defined by a list of conditions necessary and sufficient for class membership.[3] Then the features shared by the referents of a term (the members of the corresponding category) would divide into two disjoint sets, those shared by virtue of the definition and those shared contingently, because of the way the world actually is. A similar division would occur in the descriptive statements about members of the category. Some would be true by definition: no observations could make them less likely. Such statements the logical empiricists called *analytic*. The others—so-called synthetic statements—were to be evaluated on the basis of observation. Analytic truths were thus linguistic, the tautological consequences of accepted social conventions. Synthetic statements were empirical, to be believed or rejected on the basis of evidence.

A standard philosophical example will indicate both the nature of the analytic/synthetic distinction and the difficulties it poses. Books on philosophy of science have, at least until recently, frequently used statements like "All swans are white" (or "All ravens, crows, rooks, etc. are black") to illustrate the nature of nomic or lawlike generalizations, the truth of which is to be judged on the basis of neutral observation. These books emphasized that the observation of many swans, all of them white, increases the plausibility of such a statement, but they stress also that no amount of observation could prove the generalization true: a swan of another color might still be found at another place or time. If it were found, the argument continued, if a single nonwhite swan were observed, the statement would be proved false. That asymmetry between truth and falsification was long held to characterize all universal empirical generalization. Unlike

_____

vehement debate currently surrounding them to make some needed points about the development of the sciences more generally.

3. This discussion treats explicitly only the traditional theories which take meaning to be given by a set of necessary and sufficient conditions. But it is readily adapted also to the looser, recently more prevalent theories that talk, not of necessary conditions, but of a cluster of conditions of which a certain number or percentage must be satisfied in order that the term in question apply. The key difficulty for these so-called cluster theories is to provide a basis for specifying the number or percentage of cluster members required to specify reference.

definitions, which were analytic conventions, immune to experience, no empirical generalization was necessarily true; synthetic statements were in principle corrigible.

Such discussions all required the now widely questioned assumption that the people who gathered evidence about the color of swans shared a definition of *swan*, a way of picking out swans, that was independent of beliefs about their properties, in this case about their universal whiteness. But no such definitions were ever supplied for cases of this sort, a lacuna referred to in chapter 1 as the failure of attempts to supply a neutral, belief-independent observation language. The present example suggests the source of the difficulty. My dictionary tells me that swans are "mostly pure white aquatic birds," and whiteness is certainly among the characteristics I lean on heavily in identifying them. If early Pacific voyagers had not discovered in Australia black waterfowl otherwise just like swans, my dictionary would presumably have omitted the qualifier *mostly*. In that case, whiteness would have been available as a defining feature of *swan*.

A definition of *swan* that included whiteness would, however, put its users at risk. Finding a black waterfowl which closely resembled a swan in all characteristics except color, they would be forced to choose between two awkward alternatives. The first would require them to seek a new definition, acknowledging that the one previously in use had not, because a product of belief, been properly a definition at all. Their mistake was not simply about the world, but about the logic of definition as well. The second would require the conclusion that the newly discovered swanlike creature was not, by incorrigible definition, a swan, but some new sort of creature. On this second scenario, convention would have deprived the definition's users of information that might otherwise have increased their understanding both of swans and of the role of color in distinguishing kinds of fowl. A similar, if more far-fetched, scenario can be devoted to the choice of *aquatic* as a defining feature. No nontheoretical (i.e., belief-independent) reason bars the discovery in some previously unexplored territory of a group of snow-white, swanlike creatures which never go near water except to drink. In either of these cases, the natural thing to do when confronted with unexpected swanlike creatures is to keep options open. Only further experience with the anomalous creatures can provide a basis for deciding whether they are swans or members of a previously unknown kind.

Analyses of this sort, many of them far more rigorous and detailed, have, in the last half century, led philosophers to the perhaps obvious conclusion: that a line between analytic and synthetic statements cannot ordinarily be drawn. The intension of a term, the characteristics that enable a user to pick out the term's referents, are themselves products of experience and thus dependent on beliefs about the way those referents behave. With a few debatable exceptions, most of them in logic and mathematics, no statements are true by virtue of meaning alone; all are subject to correction on the basis of further experience. Whatever certainty science may possess cannot be traced to the certainty of the individual statements which embody its content. This was the point stressed by W. V. O. Quine in the classic essay which announced the demise of the analytic/synthetic distinction: "Taken collectively," he wrote, "science has its double dependence on language and experience; but this duality is not significantly traceable into the statements of science taken one by one." A widely quoted metaphor pinned down the point a few lines later: "The totality of our so-called knowledge or beliefs . . . is a man-made fabric which impinges on experience only along the edges."[4]

Though I shall later question whether so extreme a holism is required—whether "the totality of knowledge or beliefs" cannot be divided into localizable blocks for purpose of analysis—these remarks of Quine's seem to me just right. They have, furthermore, brought needed liberation to a tradition in philosophy of science for which the analytic/synthetic distinction had proved a procrustean constraint. But it is important to recognize that, as Quine's phrase "double dependence" suggests, the removal of the distinction has two aspects which need, so far as possible, to be separated. The first is a threat to a neutral foundation for science, a foundation held to be independent of language and culture. The second is a threat to a long-standing understanding of meaning, in particular to the distinction between the intension and extension of words and phrases. These two aspects are closely interrelated, as their history and the argument of this book

---

4. W. V. O. Quine, "Two Dogmas of Empiricism," in *From a Logical Point of View* (Cambridge, MA: Harvard University Press, 1953), 42. The essay was originally published in 1951, and I am one of the many greatly indebted to it. [The original publication was in the *Philosophical Review* 60 no. 1 (1951): 20–43.]

should suggest.[5] But seeing them as two makes it easier to recognize the alternative paths which the demise of the analytic/synthetic distinction opens.

One is the path followed by Quine and many other analytic philosophers. They preserve a neutral, culture-free foundation for knowledge, but to do so they abandon the notion of meaning or intension. Words, phrases, and statements are to be understood extensionally, in terms of their referents alone. What a word means is simply the set of things to which it refers. This book takes the other fork, following the developmental path indicated in chapter 1. It abandons, that is, the long-standing assumption that a neutral, culture-free foundation is required in order to validate the cognitive achievements of science: a localized, historically situated foundation will, it argues, entirely suffice. On the other hand, making a case for that position requires the rehabilitation of a conception of meaning that includes more than extension. A developmental explanation of science needs an account of the way in which the knowledge and beliefs of one generation are transmitted to their successors for further development. That account, in turn, requires a distinction between learning what a term means (something which is discovered during the transmission process) and learning something previously unknown about the things or situations to which that term refers.[6] If the intension of a word or phrase cannot be understood as the necessary and sufficient condition for its application, then some substitute must be supplied. In the event, the element that provides that substitute will prove to be the one that gave coherence to the examples presented in chapter 2, transforming them from a bare list of the referents of terms like *motion, place, couple,* and *energy element* to portions of an integrated system of beliefs. Sketching a notion of meaning capable of supporting such points is the aim of part II, below.[b] The remainder of this chapter continues preparation for it.

5. For their interrelationship, see again Hacking, *Why does Language Matter to Philosophy?*

6. The so-called causal theory of reference attempts to provide such a distinction in purely extensional terms, but it cannot be adapted to cases like the ones discussed in chapter 2. For the nature of its inadequacy, see my "Possible Worlds in History of Science," in *Possible Worlds in Humanities, Arts, and Sciences: Proceedings of Nobel Symposium 65*, ed. Sture Allén (Berlin: Walter de Gruyter, 1989), 9–32 [reprinted as chap. 3 in *Road Since Structure*].

II

To the required reformulation of the concept of meaning or intension, the case histories in the preceding chapter provide two important clues. The first, mentioned at the start of this chapter, is that the terms that require reinterpretation in order to remove anomaly and gain understanding are all of them kind terms. They are, that is, terms that refer to the sorts of objects, materials, situations, or properties which could occur in the natural or social world. In the examples they included *motion*, *body*, *property*, and *place*; *cell*, *current*, *liquid*, and *resistance*; *energy*, *quantum*, *radiation*, and *resonator*. In the social world they would include *astronomer*, *democracy*, *negotiation*, and *teacher*. All such terms are syntactically labeled. In English and most Romance languages they take the indefinite article, either by themselves (e.g., "*a* motion," "*a* resonator") or, in the case of some mass terms, when coupled with a qualifier, a term or phrase, that does (e.g., "*a* gold ring" or "*a* cup of water"). Other languages label kind terms in other ways, but some sort of label for them seems to be universal. Part of learning a kind term is recognizing its label, learning that it is and behaves like a kind term. That is, if you will, part of learning what the term means.[7]

The second clue was stressed when presenting the examples of the last chapter. The kind terms and kind concepts requiring reinterpretation are regularly interrelated in localized groups that must change together if a coherent reading is to be retained. Consider *motion*, *matter*, and *property* in their Aristotelian versus their Newtonian senses; Volta's *couple*, *discharge*, and *leakage* versus the subsequent *cell*, *current*, and *resistance*; or Planck's original *element* and *resonator* versus his later *quantum* and *oscillator*.

---

7. I am restricting attention to nouns, the most familiar and, for present purposes, the most important case. But there are some adjectival kind terms, especially but not only those from which nouns are derived or which are derived from them: the adjectival and nominal use of *male* is one example, the pair *carnivorous* and *carnivore* provides another. There are also verbs that show the pattern of kind terms, for example the gaits of a horse: *walk*, *trot*, *canter*, and *gallop*. Extension of the following discussion to grammatical classes other than nouns is a clear desideratum, but my present case does not depend on its achievement, and I shall bypass the complex issues it presents. On this question, see also Eli Hirsch, *The Concept of Identity* (New York: Oxford University Press, 1982), esp. 38. Hirsch, many of whose problems are the same as mine, adopts the same policy with respect to adjectival and verbal kind terms. (See also note 8, below.)

These two clues prove to be inseparable, and the path to be followed in pursuing them will be long. It leads first, in the three chapters of part II, to a replacement theory of meaning for kind terms, a theory which in turn opens the way to the problems of progress, relativism, and truth, discussed in the three chapters of part III. Before starting upon it, I shall in the remainder of this chapter anticipate some of the main conclusions that will emerge along the way, especially those relevant to the replacement theory of meaning. Marking the trail in this way may make it easier to travel and perhaps also motivate its pursuit.

Kind terms prove to be of two sorts, which I shall label *taxonomic kinds* and *singletons*. Both have nonlinguistic predecessors, and both play fundamental roles in the language of everyday life. Taxonomic kinds are the more familiar and in the everyday vocabulary overwhelmingly the more numerous. Like individual organisms, the members of a taxonomic kind are members of a species within which they are distinguished by their individual differences. That species in turn belongs, with a few others, to a higher-level category (for reasons to appear, I shall be calling it a *contrast set*) within which species membership can be determined by features that discriminate between members of the different species within the set. Biological kinds are taxonomic, as are all of the social kinds mentioned in the first paragraph of this section: *astronomer* is in the genus that includes *chemist, physicist, geologist,* and others; *democracy* in the contrast set that includes *monarchy, autocracy, dictatorship*; and *negotiation* in the set that includes *arbitration, mediation,* and so on.

In the examples from chapter 2, taxonomic kinds are far scarcer, a significant difference about which more will shortly be said. But there are a few, most obviously in the Aristotle example, which deals with everyday phenomena and uses a vocabulary closer to everyday life than that of the other examples. The subcategories of change form a contrast set, as do the beginning and end points of a change. The four Aristotelian elements are another example, this one with a continuous history into and through the invention and explanation of the periodic table. Still another ancient set of taxonomic kinds with a continuous history in the sciences is the population of the heavens. In Aristotle and other ancient writers, it was constituted in its entirety by three species: two of them corresponded roughly to our stars and planets, the latter, however, including the Sun and Moon; the third species, meteors, contained and gave conceptual unity to all other celestial phenomena, including what we call comets but also rainbows and the Milky Way.

Two characteristics of taxonomic kinds are fundamental to the argument of this book. The first was described at the start of this section: the meaning of a taxonomic kind term is bound up with the meanings of the other kind terms in the same set; none has meaning independent of the others. Second, the ability to identify members of a taxonomic kind can be acquired by direct exposure to members of that kind and of the other kinds in its set. Though some signal is needed to tell the learner whether a given attempt at identification is right or wrong, no words are required. That is why evolution has equipped animals to learn to discriminate the variety of kinds basic to survival in their environments.[8]

Singletons are very different. They are especially prominent in the sciences where they play a central role. Of the kind terms drawn illustratively from the examples of chapter 2 at the beginning of this section—*motion, body, property*, and *place; cell, current, liquid*, and *resistance; energy, quantum, radiation*, and *resonator*—all but *liquid* (a descendent of an Aristotelian taxonomic kind, the element water) are singletons, not members of any contrast set. Few singletons appear in the everyday vocabulary, and the concepts for which those few supply names are the common possession of man and at least the higher animals. In societies with language, they are fundamental categories of thought: space, time, and individuated physical body; perhaps also a root concept of cause and of such basic social categories as self and other. They appear, collectively, to be innate, and chapter 4 will suggest that their likely evolutionary source is the neural processes developed to track moving objects and to match different situations to a repertoire of behavioral responses. Chapter 5 will show how that source can generate taxonomic kinds.

---

8. If the kinds involved are social kinds and the society in question has a language, then the learner may need to understand what the members of the relevant kind—astronomers, say, or the parties to a negotiation—are saying to each other. But they need not in principle be told anything at all about what terms like *astronomer* or *negotiation* mean. Extra information of that sort can ease the process of interpretation, but it can be carried through without linguistic guidance.

As either their name or the preceding examples suggest, singleton kinds are not grouped with somewhat similar kinds in contrast sets but are instead sui generis. Nevertheless, as the example of space, time, and body may suggest, they are interdependent. Like taxonomic kinds, they must be learned together in small local groups. But, excepting the original innate group, they cannot be learned without language, merely by pointing to examples. Instead, as will be shown in chapter 6, two or more of them must ordinarily be grouped together in sentences, and examples must then be given, in words or by the exhibit, of situations to which those sentences apply. These sentences are not definitions of the terms learned with their aid, but they have an apparently universal necessity, like the one Kant labeled the *synthetic a priori*. Many of them are described in the sciences as laws of nature, and the terms learned with them resemble those traditionally labeled *theoretical terms*. Hooke's law and Newton's second law of motion will provide examples of the lawlike sentences; *force* and *mass* of the terms acquired with them. The discovery of additional generalizations like them marks the progress of science.

Though singletons increasingly dominate the vocabulary characteristic of the sciences, taxonomic kinds do not vanish but instead are largely driven underground. And even that does not occur entirely. There are important taxonomic or natural historical sciences. In addition, as my earlier references to kinds with continuous histories from antiquity should suggest, the introduction of new singletons occasionally permits more refined analysis and sometimes also explanation of taxonomic kinds already in place, e.g., the arrangement of elements in the periodic table. And finally, new singletons sometimes make possible the introduction and study of new taxonomic kinds, like the fundamental particles of modern physics. But these examples are, for present purposes, secondary. The overwhelmingly greatest importance of taxonomic kinds to the sciences is their role in the experimental and instrumental practices essential to scientific development.

Sometimes those practices bring singletons and their accompanying laws into being, and sometimes, conversely, those practices are brought into being by the introduction of new singletons. But once in being, those practices can and usually do have a life of their own, a life to be discussed when singletons are considered in chapter 6. They can, that is, be carried forward by people who do not control the singletons, laws, and theories which scientific development has associated with them. And conversely, the research reports through which science advances can be silent about instrumental practices, describing an instrument only when it is of a new sort and itself still a subject of study. (Note the absence from the second and third examples of even the names of the instruments that made the research possible, instruments like electroscopes, galvanometers, experimental cavities, and bolometers.) The established, quasi-independent instrumental practices which underlie all research are taken for granted in those reports, together with eyes, ears, and hands, the primitive instruments which these more developed instruments extend. Taken for granted also is the language in which the results of instrumental practices are recorded.

III

Both singletons and taxonomic kinds are subject to an important prohibition which I shall, in the pages to follow, call the *no-overlap principle*. Everyday biological kinds suggest its nature. Children learning to use the term *dog* must ordinarily learn the use of *cat* as well. In environments where both kinds are present, the two terms belong in the same contrast set and must then be acquired together. But both the success and the utility of that acquisition require that the world contain no dogs which are also cats and vice versa. The everyday species dog must not, that is, overlap the species cat, though both are entirely contained within the class of animals and both overlap the class of four-footed animals. For taxonomic kinds the no-overlap principle requires that the kinds within a single contrast set be entirely disjoint, share no members. Chapter 5 will suggest that, for success in practice, a slightly weaker form of the principle will suffice. Encounters with creatures that seem candidates for membership in two or more kinds within the same contrast set must be exceedingly rare. Frequent difficulties in deciding whether some recently encountered animals are, in fact, dogs or cats indicate a failure of the category system.[9]

9. For present purposes, satisfaction of the no-overlap principle is the necessary and sufficient condition for being a kind term, and it can play that role also for the exemplary adjectives and the verbs discussed in note 7, above: *carnivorous* vs. *herbivorous*, for example, or *canter* vs. *trot*. But what is one to say about *red* and *blue*, used adjectivally? I would like to declare that a body may be red or blue, reddish-blue or bluish-red, but not red-blue. The concluding

For singletons the principle is stronger, amounting to a form of the principle of noncontradiction. Singletons must be acquired, I have suggested, together with universal, lawlike generalizations. For them the no-overlap principle states simply that no two incompatible generalizations can apply to the same particular member of a singleton.ᶜ Just as there can be no dog which is also a cat, so there can be no force which satisfies both Newton's second law of motion and some other law incompatible with it. In chapter 9 I shall suggest that the laws of noncontradiction and of the excluded middle are articulations of the no-overlap principle for singletons and for taxonomic kinds, respectively.

This description of the no-overlap principle may seem to confirm a confusion that philosophically adept readers will already have suspected. Is it the world or language to which the no-overlap principle applies? Am I talking about terms, the kind term *cat*, for example, or about the things to which the term applies, my own cat Gertrude, for example? To those questions and others like them the answer is: I am talking about both, and it is to both that the no-overlap principle applies![10] Much additional time and space may be required to make that response plausible, but it is not too early to begin.

The problem of the analytic/synthetic distinction, discussed above as a problem about definitions, is a reflection of a deep and indissoluble entanglement between the kind terms of a language and the world that users of that language inhabit. That language is an inheritance. Before it was acquired by its present users, successive generations tuned and polished it to fit the world to which their successive versions of the language gave access. As a result of such tuning, the members of each generation inherit an instrument superbly adapted to their own natural and social environment. But the efficacy of that instrument is bought at the expense of its

---

prohibition is, however, dubious, and it illustrates the central reason why I've skirted the issue of kind terms other than nouns. The issues they raise are, I take it, among those that Wittgenstein is exploring in *Remarks on Colour*, ed. G. E. M. Anscombe, trans. Linda L. McAlister and Margarete Schättle (Berkeley: University of California Press, 1977).

10. One could introduce two no-overlap principles at this point, one for terms and the other for their referents. I have treated them as one, because each turns out to be an inescapable consequence of the other and neither has ontological priority. That treatment need not suggest that there is no distinction between words and things, though it is often taken to do so.

universality, and that limitation occasionally shows. On the one hand, the kind terms deployed in the language at a given time permit its users to deal as precisely as circumstances require with the expected, with the range of phenomena, that is, to which the language has been tuned. In addition, the roster of accepted kind terms can often be enriched in the face of new situations: foxes and wolves can be added as kinds to the set that, as originally acquired, contained only dogs and cats. But the roster of kind terms cannot always be enriched to deal with novelty. The addition of terms which violate the no-overlap principle is barred: the discovery of a community of dog-cats or of a non-Newtonian force law requires not an addition to but a revision both of the categories exemplified by natural phenomena and of the vocabulary in which those phenomena are described. Such revisions occur frequently in scientific development: all the examples in the previous chapter provide examples.

Readers of *Structure* may by now hear echoes of a fundamental theme closely associated in that book with the notion of paradigm: a fundamental tool presupposed by the members of a group in their dealings with each other and with their world, [which] limits what those dealings can accomplish. Here, that tool is the arrangement of kind terms that I am calling a *structured kind set*. Its acquisition is a precondition for membership in the community of its users, and that acquisition takes place through a socialization process which equips neophytes with a set of interrelated kinds together with some minimal knowledge about them, some expectations, that is, about the ways the members of each kind behave. Only with that equipment in place is the neophyte prepared to engage in community practice and the discourse central to it.

What it is for two or more individuals to share a kind set will be developed in the next three chapters. Here I note only that it does not require sharing all the same beliefs and expectations about the members of the set but that it nevertheless, by virtue of the no-overlap principle, greatly restricts what those beliefs may be. That restriction is crucial, and I know no better way to introduce it. But this way, unfortunately, often misleads. Its point is not that the acquisition of a particular kind set bars *belief* in certain proposition[s] (though in a Pickwickian sense, it does). Rather, the acquisition of a particular kind set bars *even the formulation*, conceptual or

verbal, of certain beliefs held by users of another kind set. Different or differently structured kind sets give access to different, though largely overlapping, ranges of possible belief. Users of one kind set must be able to
suspend or bracket it in order to gain access to some of the propositions
that are candidates for belief to the users of the other kind set.[11]

IV

The examples of the last chapter were intended, first and foremost, to illustrate the process and consequences of gaining that access, and the notion
of a kind set suggests a way to reformulate their significance. Attempting to
read a text from an earlier time, the historian encounters anomalies which
isolate one or more local disparities between the kind set he or she has
brought to the text and that of the person who composed it. Recognition of
such a disparity inaugurates an attempt to recover the apparently anomalous kinds used by the author. That attempt begins with the collection of
passages in which one or another of the troublesome terms occurs. It continues with attempts to discover the common characteristics of the various
occasions on which these terms are used, whether separately or together,
and to discover simultaneously what differentiates those occasions from
occasions we would treat as similar but which the author does not. These
techniques of collection and analysis underlay the examples presented in
the last chapter, though they rarely surfaced explicitly. Except by example,
such techniques are difficult to teach, as I wrote at the beginning of chapter 2,
but they do leave informative traces when applied. Look again, for example, at both the data on and the discussion of *chora* and *topos* in footnote 28
in chapter 2, and note also, early in that chapter, the discussion of the characteristics which unite all sorts of change under the category *kinesis*; or
consider the account of the use of *quanta* and *element* in Planck's papers
and elsewhere in German physics.

---

11. The phrase "candidates for belief" is modeled on Ian Hacking's "candidates for truth
or falsehood," which I take to have identical intent. In any case, I've been much assisted by the
discussion in which he introduced the phrase: "Language, Truth, and Reason," in *Rationality
and Relativism*, ed. Martin Hollis and Steven Lukes (Cambridge, MA: MIT Press, 1982), 48–66.

The techniques and experiences I have attributed to my imagined historian struggling to break into a text are close to those Quine attributes to a figure he describes as a "radical translator," an imaginary anthropologist struggling to learn the natives' language by observation of their behavior. Both Quine and I use the term *interpretation* to describe the process, and a comparison of our views of its result proves illuminating. About three of its most important characteristics we agree fully. First, for the interpreter, whether my historian or Quine's anthropologist, the unit of analysis is always, at the least, a full sentence or statement. Though the anomaly that requires interpretation may be signaled by the use of a single word, it is not the word itself but its use in that sentence that is anomalous. Whole statements must therefore be the primary bearers of meaning.[12] Second, the process of interpretation is never finished and may at any time go astray: the next sentence or the next native utterance may prove anomalous, requiring extension, refinement, or even reform of what has gone before. Finally, some hypotheses (Quine calls them "analytic hypotheses") about native or authorial behavior nevertheless work much better than others: what was anomalous becomes part of a new order; what was incomprehensible before can now be understood. Though the interpreter, historian, or anthropologist, must always be wary, on the alert for new surprises, there is good reason meanwhile to rely on the understanding that interpretation has already achieved.

The steps so far described do not, however, complete the interpreter's task. The results of interpretation must still be communicated to an audience in the interpreter's own culture, and it is with respect to how this is done that Quine's and my paths diverge.[13] The rest of this section discusses his; mine will be sketched in the section to follow, the last in this chapter.

12. These statements or sentences need not, of course, satisfy the grammatical criteria of the developed language. *Mama* or *doggie* uttered by a child in the presence of a potentially appropriate object counts as a sentence.

13. Discovering the nature of that divergence has been crucial to the development of the views presented in this book, and the challenge which resulted in that discovery is perhaps my principal debt to Quine. That debt dates from the academic year 1958–59. He and I were both at the Center for Advanced Study in the Behavioral Sciences [at Stanford University], and he circulated a draft of chapter 2 of *Word and Object* (Cambridge, MA: MIT Press, 1960). References

For Quine the anthropologist-interpreter is a translator who transmits native linguistic behavior to his or her audience in that audience's own language. Taking that to be the case, Quine further argues that for most statements in the natives' language an infinite number of entirely equivalent translations are possible, that the proper result of translation is fundamentally indeterminate. For simplicity Quine begins by considering homophonic translation, the mapping of the sentences of a language onto sentences of the same language, and he concludes as follows: "The infinite totality of sentences of any given speaker's language can be so permuted, or mapped onto itself, that (a) the totality of the speaker's dispositions to verbal behavior remains invariant, and yet (b) the mapping is no mere correlation of sentences with *equivalent* sentences, in any plausible sense of equivalence however loose. Sentences without number can diverge drastically from their respective correlates, yet the divergences can systematically so offset one another that the overall pattern of associations of sentences with one another and with non-verbal stimulation is preserved."[14] What holds for mappings within a language must hold for mappings between languages as well. The result is what Quine calls the *indeterminacy of translation.*

For me, Quine's conclusion seems a reductio ad absurdum of his argument, and one of its premises seems clearly responsible. Quine still seeks a fixed Archimedean platform, and he retrieves it by premising a privileged class of sentences, which can be evaluated on the basis of sensory stimulation alone and which thus "sustain the philosophical doctrine of the infallibility of observation sentences."[15] Language acquisition begins with such observation sentences; they provide the foundation required to gain command of sentences of a more elaborate sort; and they are all for Quine *transcultural,* in the sense that they can be translated fully into the language

---

in my work to Quine on translation begin a few years later but remain equivocal and occasionally contradictory until the early 1980s.

14. Quine, *Word and Object*, 27. A helpful discussion of what this mapping amounts to is provided by Hilary Putnam in *Reason, Truth, and History* (Cambridge: Cambridge University Press, 1981), 32–38, 217ff. But [Putnam's] example should be read with caution, for it deals only with a simple case, a set of observation sentences, and what results from his mapping *is* a "mere correlation of sentences with *equivalent* sentences." The indeterminacy it illustrates is trivial, but that, we are about to see, is what Quine would have expected.

15. Quine, *Word and Object*, 44.

of any other culture.[16] The observation sentences of any pair of languages can thus be mapped onto each other in only one way, reducing them in effect to a single language. Indeterminacy of translation is encountered only with more elaborate sentences, and the argument that it is encountered even there requires as a premise the neutrality of observation sentences.

That premise has, of course, been deeply ingrained in the empiricist tradition since the seventeenth century, when its origin was closely associated with that of empirical science.[17] But it is nonetheless mistaken, as any translator knows. *Traduttore traditore* applies to observation sentences at least as well as to the more elaborate sorts of sentences to which they lead.[d] Eugene Nida, a prominent contributor to the literature on translation, writes: "*All types of translation* involve (1) loss of information, (2) addition of information, and/or (3) skewing of information."[18] An especially telling example of all three is provided by the linguist John Lyons. He considers the English sentence "The cat sits on the mat," asks how it is to be translated into French,

16. For the translatability of observation sentences, see Quine, *Word and Object*, 68. I am deeply indebted to James Conant for calling to my attention the centrality of this latter passage to my argument.

17. Francis Bacon denounced the language of common speech—his "idols of the market"—for its tendency to mislead [see the *Novum Organon*, book 1, in *The Works of Francis Bacon*, ed. James Spedding, Robert Leslie Ellis, and Douglas Denon Heath (New York: Garret Press, 1968)]; the Royal Society adopted *nullius in verba* [usually translated as "take no one's mere word as evidence"] as its motto; and important proponents of the new philosophy took up the widespread search for a universal character, a plain language into which all known forms of speech could be translated. On this whole subject see M. M. Slaughter, *Universal Languages and Scientific Taxonomy in the Seventeenth Century* (Cambridge: Cambridge University Press, 1982). See also note 4, above.

18. Eugene Nida, "Principles of Translation as Exemplified by Bible Translating," in *On Translation*, ed. Reuben A. Brower (Cambridge, MA: Harvard University Press, 1959), 13 (italics mine). Like much of the other writing on the subject, this volume deals primarily with translation of literature. But it contains much of use—especially the brief essay (232–39) by Roman Jakobson, "On Linguistic Aspects of Translating"—including its bibliography. Walter Benjamin's "The Task of the Translator," originally published in 1923 as a preface to his translation of [Charles] Baudelaire's *Tableaux Parisiens* [Heidelberg: Verlag von Richard Weissbach], is an especially penetrating study of the problems of translation and of the extent to which they can and cannot be resolved. It is easily available in his collection *Illuminations*, ed. Hannah Arendt, trans. Harry Zohn (New York: Harcourt Brace & World, 1968). George Steiner's widely known *After Babel: Aspects of Language and Translation* (New York: Oxford University Press, 1975) includes a wide-ranging selected bibliography and an interesting personal account of the problems and rewards of multilingualism.

and concludes that, in any strict sense, it cannot be.[19] After noting the differences in reference between English *cat* and French *chat* and noting that French grammar forces the translator of *sits* to choose between the act of sitting and the state of being seated, Lyons continues as follows:

> The translation of "the mat" is more interesting. Is it a door-mat that is being referred to ("paillasson"), or a bedside mat ("descente de lit"), or a small rug ("tapis")—not to mention various other possibilities? There is a set of lexemes in English, "mat," "rug," "carpet," etc., and a set of lexemes in French[,] "tapis," "paillasson," "carpette," etc.; and none of the French words has the same denotation [reference] as any one of the English lexemes. Each set of lexemes divides, or categorizes, a certain part of the universe of domestic furnishings in a different way; and the two systems of categorization are incommensurate . . .
>
> It is only too easy to be aware of the difficulties in translating from one language into another and yet to underestimate, or miss completely, the theoretical implications of the facts which give rise to these difficulties . . . The denotation of a lexeme is limited by the relations of sense which hold between it and other lexemes in the same language. The denotation of "mat" is limited by its contrast in sense with "rug" and "carpet"; the denotation of "paillasson" in French is limited by its contrast in sense with "tapis" and other lexemes. We could not reasonably say that "mat" has two meanings because it is translatable into French by means of two non-synonymous lexemes, "tapis" and "paillasson"; or that "tapis" has three meanings because it can be translated into English with three non-synonymous lexemes "rug," "carpet," and "mat." The meanings of words (their sense and denotation) are internal to the language to which they belong.

Note now that "The cat sits on the mat" is an observation sentence, the occasions for its linguistically appropriate utterance being determinable

---

19. John Lyons, *Semantics*, vol. 1 (Cambridge: Cambridge University Press, 1984), 238. Among Jehane Kuhn's notable contributions to this book, the gift of this passage from Lyons is not the least.

by sensory stimulation alone.[20] Lyons's discussion invites us to imagine an English-speaking guide exhibiting to a French-speaking visitor a series of situations which could appropriately lead English speakers to say, "The cat sits on the mat." For each of these situations, the visitor would have an appropriate French utterance, and the utterance would be an observation sentence too. But the appropriate French observation sentence would vary from one situation to another. No single French sentence would translate all the appropriate utterances of "The cat sits on the mat." Nor could there be any universal French generalization that applies to all and only situations in which a cat sits on a mat. For practical purposes (buying a new floor covering, for example), that untranslatability may be unimportant. But universal generalizations are constitutive of much of science, and their untranslatability can there be deeply consequential.

This problem of translation goes deeper. When, in a particular situation, the English speaker says, "The cat sits on the mat," and the French speaker utters the corresponding French observation sentence, the French sentence does not translate the English and vice versa. Though both sentences apply to the particular situation that has evoked it, there are countless other situations to which one would apply, the other not. Enriching French by addition of the term *mat* will not serve the purpose either. Like *paillasson, tapis,* etc., *mat* is a kind term. Its meaning, whatever that may be, depends on its presence in a contrast set that contains such nonoverlapping English kind terms for floor coverings as *carpet, rug,* and so on. If it could be imported into the French category for floor coverings, it would be known through different contrasts and would have a different meaning. Those differences might be tolerable if members of the new category were entirely distinct from the members of those already in place. But the addition of *mat* to the French category of floor coverings violates the no-overlap principle. *Mat* and *paillasson,* for example, overlap: they share some but not all referents. Only one of the two can therefore bear the label that identifies it as a kind term and that is therefore essential to its functioning. Though languages do get enriched by borrowing and in other ways besides,

---

20. Lyons actually considers the sentence "The cat sat on the mat." I have changed *sat* to *sit* to eliminate the role of memory in its evaluation.

the enrichment takes time, and it results in more than a mere addition to what was there before. Application of the term *enrichment* to linguistic development is problematic in the same way as application of the term *growth* to scientific development. Indeed, as the examples may have suggested, the two sets of problems are the same.

None of what precedes is intended to deny either the existence or the importance of observation sentences. Nor should it suggest that observation sentences cannot frequently be translated from one language to another. The point is rather that the body of observation sentences of one language seldom if ever correlates one to one with the observations sentences of another. Furthermore, the only way to find out which ones do and which do not correlate is by living with native speakers and sharpening one's sensitivities to anomalous bits of text or of behavior. There are no criteria outside the pair of languages being compared to determine which sentences will be translatable, which not.

V

One last aspect of the passage from Lyons is also relevant. It uses the word *incommensurate* to describe the relation between the English and French ways of categorizing floor coverings, and in doing so it introduces what is arguably the fundamental concept of this book. Over thirty years ago, Paul Feyerabend and I borrowed that term's near relative, *incommensurability*, to describe the relationship between an older and a more recent scientific theory.[21] Both forms are, of course, drawn from mathematics, where they mean no common measure (the standard example is the relation between the side and diagonal of an isosceles right triangle). In their borrowed use they mean no common language, no universal character, into which all sentences from both linguistically expressed theories can be translated. Together with the examples of chapter 2, this chapter has aimed to show the appropriateness of the metaphor. Translation is at best an imperfect bridge

---

21. I have said a bit about our apparently independent introduction of the term in my "Commensurability, Comparability, Communicability," *PSA 1982*, vol. 2 (East Lansing, MI: Philosophy of Science Association, 1983) [reprinted as chap. 2 in *Road Since Structure*].

to the thought of another culture or of an earlier age: incommensurability of kind concepts bars its full use.

Where translation fails, however, another recourse is available. Faced with anomalous passages that suggest incommensurability, one can, by the process Quine and I label *interpretation*, attempt to learn the language in which those passages were written. Not translation but language learning is the process in which his imagined anthropologist and my imagined historian were engaged. Though it is prerequisite to the actual practice of translation, what it produces in the first instance is, at best, bilinguals. They can understand both languages and respond to what is said in them. But what they hear and what they say cannot always be expressed in both languages, and they must therefore constantly be aware in which language community they are participating. Bilinguals can, that is, move from culture to culture, but their behavior must here and there change as they do so. Failure to make such changes results in behavior [being] received as anomalous.

Think now of the examples in chapter 2. Because they and their like appear to have been presented in the ordinary English brought to them by their readers, incommensurability has been called self-refuting. "To tell us that Galileo had 'incommensurable' notions," one critic writes, "*and then go on to describe them at length is totally incoherent.*"[22] But remarks of that sort miss the way in which the examples were communicated. They were, of course, presented in ordinary English, where that could be done without distortion. But in the occasional passages which communication in ordinary English made anomalous, it was necessary for the historian (in these cases mostly me) to learn the conceptual vocabulary of the author of the text, teach it to readers, and then *use it* to present the views under discussion. When translation fails, no other recourse is available. The tower of Babel will not be rebuilt: neither language communities nor cultures can be merged without impoverishing loss. But learning and teaching the other

22. Putnam, *Reason, Truth, and History*, 115 (italics Putnam's). The article cited in the previous note was primarily an attempt to reply [to Putnam]. Feyerabend replied independently but in almost identical terms, in "Putnam on Incommensurability," *British Journal for the Philosophy of Science* 38, no. 1 (1987): 75–92. In referring to Galileo, Putnam is making use of an example of Feyerabend's, but the discussion is explicitly concerned with my work as well.

language provides a powerful alternative, and, unlike translation, recourse to it is likely always to be available.

One cannot, of course, be certain of the universal accessibility of human languages to human beings. But the shared biological and environmental heritage of human beings makes universal access likely; experience suggests that it exists; and the theory of kinds to be developed in chapter 5 will increase that likelihood further. Indeed, it is hard to see how universality could fail. What would we conclude if we discovered a strange tribe, imputed language to it, and then found [that] we could not, after much effort by skilled people, acquire the language? Perhaps simply that we had not been clever enough, that more work was needed. Or perhaps that we were mistaken to attribute language to the tribe. Or perhaps instead that tribe members, though superficially like *Homo sapiens*, were really not humans. Where among these alternatives should we place the conclusion that an inaccessible human language has been found? And what criteria would we use to distinguish between them? The philosopher who wrote, "The common behavior of mankind is the system of reference by means of which we interpret an unknown language," also wrote, "If a lion could talk, we could not understand him."[23]

February 28, 1995

---

23. Ludwig Wittgenstein, *Philosophical Investigations*. I should acknowledge that the passages come far apart and that the context of the latter may make this use of it strained. [Kuhn's original reference was to the now outdated 1953 edition of *Philosophical Investigations*. In the edition that is now in use—the 4th, ed. P. M. S. Hacker and Joachim Schulte, trans. G. E. M. Anscombe, P. M. S. Hacker, and Joachim Schulte (Oxford: Wiley-Blackwell, 2009)—the title *Philosophical Investigations* refers only to the text formerly known as part I. The text formerly known as part II is now entitled *Philosophy of Psychology: A Fragment*. Proper references to Kuhn's quotes above are thus to paragraph 206 of *Philosophical Investigations* and to paragraph 327 of *Philosophy of Psychology: A Fragment*.—Ed.]

PART II

A WORLD OF KINDS

CHAPTER 4

Biological Prerequisites to Linguistic Description

Tracks and Situations

Turn now to the challenge presented in part I. The developmental process through which science moves ahead is always and only historically situated. To the extent that rational evaluation plays a role, the evaluations are comparisons of bodies of belief[s] actually current when the evaluations are made. Understanding that process and its direction requires philosophical analysis of extended narratives which recount a series of such comparisons. And the stage settings required for such narratives depend on a special sort of interpretation, one which provides a set of apparently familiar terms with new meaning. Neither the nature nor the significance of these interpretations can be understood in the absence of a theory of the nature of kind concepts and the meaning of kind terms, a theory that is not purely extensional but that relates these meanings to the ways in which their referents are determined. Sketching elements of such a theory is the purpose of part II, which begins here.

The route is elaborate and leads through unfamiliar territory. In the three chapters of part II, the book's apparent subject matter as well as the evidence and argument relevant to it will seem to change abruptly. The philosophical problems introduced in previous chapters will regain prominence only in part III. Meanwhile, I shall be exploring a part of the human cognitive apparatus which appears to provide a foundation appropriate to the resolution of those problems that will be developed in my closing chapters. The present chapter, based primarily on evidence from developmental psychology, examines the state of that apparatus with which humans and other animals enter the world. For the former, human infants, that examination is extended to include a fundamental change that occurs near the end of the first year of life and that appears to be associated with an early stage of language acquisition. The next chapter employs these findings as the basis for a theory of the meaning of the kind terms that underlie everyday life in one or another culture. And chapter 6 introduces a related theory for the singleton kinds that play so large a role in the sciences.

Descriptive languages depend on a close entanglement of the concept of a kind and the concept of an individual member of that kind. That entanglement is part of what makes the developed versions of those concepts what they are. Aristotle, whose conception of individual objects and their change was fundamental to the view of motion sketched in chapter 2, reflects that entanglement. Any concrete individual, like this man or that horse, was for him a *primary substance*; the kind or species to which that individual belonged was a *secondary substance*; and much of the individual's behavior was to be understood in terms of the kind to which it belonged.[1] Both these interrelated concepts [of a kind and of an individual member of that kind] serve functions vital to survival, and both have widespread roots throughout the animal world, where they are exhibited from the time of

---

1. For Aristotle, see *Categories* V. Three autobiographical fragments may indicate what brings me to substance and anticipate the direction in which my argument will move. My concern with kinds dates from *Structure*, but first became clearly visible in my "Second Thoughts on Paradigms," in 1969 (for a convenient version, see chap. 12 of *Essential Tension*). The shift from kinds to their individual members began in the mid-1970s, when I belatedly read Saul Kripke's "Naming and Necessity" in *Semantics of Natural Language*, ed. Donald Davidson and Gilbert Harman, 2nd ed. (Dordrecht: Reidel, 1972), 253–355 [Reprinted as *Naming and Necessity*, Harvard University Press, 1980]. From the start I thought it apparent that [the] causal theory provided a powerful and badly needed way to deal with the names of individuals, but I was more than dubious about its applicability to kinds. It provided, for example, a wonderfully persuasive way to trace the individual bodies we call Earth, Moon, Mercury, Venus, Sun, Saturn, and Jupiter through the conceptual upheaval known as the Copernican revolution, but it could not do the same for the kind *planet*. What it is to be a planet is simply not the same before and after Copernicus. Statements like "Ptolemy believed that planets go around the Earth but Copernicus showed that planets revolve about the Sun" are incoherent: the two occurrences of the term *planet* have different meanings and refer to different things. The event decisive for the further development of these thoughts was, however, much more recent: my reading of David Wiggins, *Sameness and Substance* (Cambridge, MA: Harvard University Press, 1980), in a seminar on natural kinds that Sylvain Bromberger and I offered in 1987.

birth. Their initial forms must be very old products of biological evolution: the neural apparatus which mediates them is likely to be subcortical.[2]

That initial form is, however, strikingly different from the form found among adult humans. Instead, the concepts of a kind and of an individual or substance seem initially to be independent. What I shall later call the *basic form* of the object concept is first exhibited in the spatial tracking of objects, whether the object traced is mother, a stranger, or some desired quarry. The corresponding basic form of the kind concept is displayed in the discrimination between situations that call for different behaviors. A young animal's response to conspecifics differs from its response to members of other species. Among the latter, its responses are further differentiated: it flees from predators, for example, but pursues prey.[3] Nothing in the logic of those two activities—pursuit of an object in view versus discrimination between different situations—requires their conceptual entanglements, and in human neonates, at least, there appears to be none. Only around the twelfth month does the more closely entwined adult version appear, and there is evidence that its development is associated with the acquisition of language. Whether or not an equivalent developmental change occurs in animals is an open question.[4] This book is restricted to human populations, and this chapter deals primarily with the prelinguistic

---

2. Note that here and in the next few pages the term *evolution* is applied to the *biological* development of the cognitive *apparatus*. Everywhere else in this book it is applied to the development of the *product* of that apparatus, itself assumed fixed. Only if the radical distinction between these two processes is kept firmly in mind can the strength of their parallels be fully appreciated.

3. For examples of the first of these functions, see Mark H. Johnson and John Morton, *Biology and Cognitive Development: The Case of Face Recognition* (Oxford: Blackwell, 1991). For examples of the second (and additional examples of the first), see Donald R. Griffin, *Animal Minds* (Chicago: University of Chicago Press, 1992); and Dorothy L. Cheney and Robert M. Seyfarth, *How Monkeys See the World* (Chicago: University of Chicago Press, 1990). None of the authors draws the line between these functions that I am advocating here.

4. I am greatly indebted to my colleague Susan Carey for this way of drawing the distinction between infant and adult forms of the relation between individuals and kinds, and that debt is only the most recent of a long series incurred since my arrival at MIT, in 1979. She [has] been my primary guide to the literature on child development, served repeatedly as critical sounding board for the ideas I drew from it, and, through her own most recent research, crucially altered my thinking on the subject. The later parts of this chapter would never have taken form without her intervention.

development of the object and kind concepts of their members. Consideration of the mature, linguistically embodied form of these concepts is reserved for the next chapter.

I

Begin with the infant's ability to track a moving object. In principle it may involve any and all of the five senses, but in humans vision and touch are the sensory modalities most centrally involved. Most of the detailed evidence about tracking in infancy concerns the visual system; and the pages that follow are largely restricted to it.[5] To engage that evidence, one must relinquish at the start a still widely circulated view of the nature of vision. Ever since the discovery of the retinal image in the early seventeenth century, it has been natural to think of the eye as a camera and of the visual process as like the interpretation of a photograph, a process in which memory and acquired knowledge play a part from the start.[6] But recent research, both psychological and neurological, demonstrates that retinal stimuli undergo a great deal of subcortical neural processing—some of it within the outer membrane of the eye itself—before use is made either of prior experience or of generalizations stored in memory. Among the products of that initial, preinterpretive processing are the simultaneous identifications of coherent patches in the visual field and of relative motion among them. In combination these two sorts of information yield the three-dimensional boundaries of the objects made accessible by the first stages of processing.[7]

5. The relative importance of the various sensory modalities varies, of course, from species to species. For dogs, the sense of smell is a principal tool for reidentification; for birds, hearing is of major importance; and for bats, hearing doubtless heads the list.

6. *Camera* is not an anachronism. The sixteenth-century invention of the camera obscura—a darkened chamber on an interior wall of which an inverted image of the scene outside was formed by light admitted through a small hole in the external wall opposite—played a major role in the discovery of the retinal image and in the development of seventeenth-century theories of vision. For a thorough study of the route to these new theories of vision and of the conceptual transformation they effected, see Alistair C. Crombie, "Mechanistic Hypotheses and the Scientific Study of Vision: Some Optical Ideas as a Background to the Invention of the Microscope," in *Historical Aspects of Microscopy*, ed. S. Bradbury and G.L'E. Turner (Cambridge: Heffer and Sons, 1967), 3–112.

7. For a primarily theoretical introduction to these issues, see David Marr, *Vision: A Computational Investigation into the Human Representation and Processing of Visual Information* (San Francisco: W. H. Freeman, 1982), and Shimon Ullman, *The Interpretation of Visual Motion*

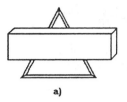 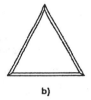 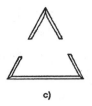

a)   b)   c)

Figure 1

Experiments suggest that these alone are sufficient to account for the observed visual-object-tracking behavior of human neonates.

Infants within the first hour after birth are regularly observed to track
moving objects by moving their eyes and turning their heads, a process apparently mediated subcortically.[8] How these neonates represent the object
of their pursuit is unknown, but by their third and fourth months experiments begin to provide clues. In one typical experiment, designed to
discover the role played by gestalt principles of well-formedness in the
perception of objects, infants of about four months were exposed to the
occluded triangle shown in figure 1a. The exposure was repeated until
the infants were habituated to the display—until, that is, the interval during which they kept their eyes on it was reduced by some predetermined
amount, often by one-half. Then half these infants were shown the display
in figure 1b, the other half that in 1c, and the interval during which they
gazed at the display was again monitored. Many experiments have shown
that, under circumstances like these, young infants will look longer at displays that seem discordant with the one to which they have been habituated. They will, that is, show more interest in a display that is novel or
surprising. In this experiment, no such preference was found. Unlike adults
and older children, four-month-old infants have no inclination to complete
the triangle. The same lack of preference between completed and divided
figures showed in numerous other experiments which adults would have

----

(Cambridge, MA: MIT Press, 1979). Views about details of the processing mechanisms discussed in these books are still in flux, but their approach and its main findings seem assured.
References to experimental explorations of the subject will be found below.

8. See, for example, Johnson and Morton, *Biology and Cognitive Development*, esp. 30–33,
78–111.

completed from clues of shape, color, or pattern. Among them was an occluded sketch of a human face.[9]

If, however, the parts of the test shapes were together in motion behind
the occluding barrier, the response of four-month-old infants was very
different. Habituated to the occluded rod in figure 2a with both rod and
block stationary, four-month-old infants showed no preference between
the completed and broken rods in the test displays b) and c). The same absence of preference was encountered if the rod and block moved together
or if the block moved and the rod remained still. But if the block stayed
still and the rod moved laterally back and forth behind it (as indicated by
the arrows in the figure), then the infants looked for more than fifty times
as long at the broken as at the complete rod. Clearly, the coherent motion
of the two exposed parts of the rod induced a strong expectation that they
were joined into a single whole behind the stationary occluder. The same
qualitative preference was shown even if the two exposed parts of the moving figure were strikingly different (e.g., a black rod on top and a broken
red hexagon below), and [it was shown] also if the rod moved vertically
or in depth rather than laterally.[10] The decisive effect of motion and the
negligible role of such qualities as shape and color in organizing the infant's
visual field into objects is striking. However disparate they may be, parts
that move together are parts of a single object.

Clearly, the infant is on the path to something like an object concept.
What can be said, on the basis of the behavior so far described, about the

9. Philip J. Kellman and Elizabeth S. Spelke, "Perception of Partly Occluded Objects in
Infancy," *Cognitive Psychology* 15, no. 4 (1983): 483–524. Useful summaries of this and related
research are given in Elizabeth S. Spelke, "Perception of Unity, Persistence, and Identity:
Thoughts on Infants' Conceptions of Objects," in *Neonate Cognition: Beyond the Blooming Buzz-
ing Confusion*, ed. Jacques Mehler and Robin Fox (Hillsdale, NJ: Elbaum, 1985), as well as in her
"Principles of Object Perception," *Cognitive Science* 14, no. 1 (1990): 29–56.

10. In addition to the references cited in note 3, see Philip J. Kellman, Elizabeth S. Spelke,
and Kenneth R. Short, "Infant Perception of Object Unity from Translatory Motion in Depth
and Vertical Translation," *Child Development* 57, no. 1 (1986): 72–86. The infant's depth perception, like its perception of lateral and vertical position, is probably monocular, the clues
once more provided by relative motion. Binocular depth-perception does not, on the average,
develop in humans until the end of the fourth month. See Richard Held, "Binocular Vision—
Behavioral and Neuronal Development," in Mehler and Fox, *Neonate Cognition*, chap. 3. More
primitive organism[s] access the depth dimension monocularly.

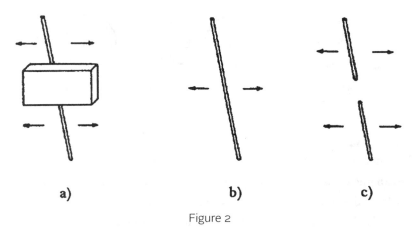

Figure 2

stage it has reached? Two points are specially relevant, the one obvious, the other less so. First, the infant behavior just described displays what I shall for now, subject to a later correction, follow [Eli] Hirsch in calling the *basic-object concept*: the object as a bounded region all of whose parts move together.[11] It is thus inextricably entangled, in a way for which the discussion of Aristotle has prepared us, with the concept of change of place and thus with some early form of the concepts which older children and adults label *space*, *time*, and *object*. Several concepts are, however, manifest together, and there is no reason to suppose that either human infants or nonhuman animals of any age separate them, have three concepts, that is, rather than one. Second, the three concepts that will later emerge are what in chapter 3 I called *singletons*, and they illustrate what I there described as the lawlike generalization that their entanglement brings with it. In this case one of those generalizations is the principle of impenetrability: no two objects can occupy the same region of space at the same time. If the neural system represents objects as bounded regions the parts of which move together, then the regions it represents cannot, in principle, both interpenetrate and remain [distinct] objects.

Figure 3 sketches a two-dimensional version of the required argument. If A and B are objects, then the neural representation in 3a is impossible.

11. For the isolation of a basic-object concept and a useful discussion of its strengths and limitations, see Eli Hirsch, *The Concept of Identity* (New York: Oxford University Press, 1982). Much of the rest of this chapter and of the next deal with the elimination of those limitations.

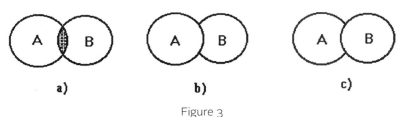

Figure 3

Points in the stippled region where the two might intersect must move with one or the other. If with A, then the required representation is 3b; if with B, then the representation is 3c. In the first, object A occludes object B (or else B is concave and fitted to A's surface); in the second, B occludes A (or A's concavity is fitted to B).

The prohibition of configurations like the one in 3a is a sort of no-overlap principle. In this form, which is purely geometric, it is the physical principle of impenetrability, and it provides a first example of a sort of the necessity I earlier likened to Kant's synthetic a priori. Though object impenetrability is a product of experience, that experience is embodied in the neural systems that result from biological evolution. For individual organisms it is prior to experiences of particular bodies, experiences for which it supplies not evidence but a precondition. Reference to the synthetic a priori calls attention, however, to an important distinction between Kant's position and the one taken here, a difference disguised by the vocabulary I have so far been obliged to use. Kant believed that the concepts prerequisite to experience—concepts like space, time, and object—were uniquely those of Newtonian physics. But the discussion of Aristotle in chapter 2 or the role of non-Euclidean geometries in modern physics indicate that alternative sets of concepts can serve the same functions. All of these sets must find a place for physical necessities like the principle of object impenetrability (look again at figure 3 and note the way it is embodied in Aristotle's definition of place), but they may otherwise differ radically. What we have so far been considering is their neurologically embedded primitive form, a form in which the three are inextricably entangled. Their disentanglement and their further transformation to one or another of their viable adult forms requires a further learning process, for which this first form is only prerequisite. Their separation is probably associated with an early stage of language acquisition, and their later development requires a fully developed syntax.

If the claim that four-month-old infants display the basic-object concept is correct, then infants of that age should show direct awareness of object impenetrability, and a variety of experiments indicate that they do. Of these the most revealing was reported by [T. G. R.] Bower in 1967 and has been repeatedly examined since.[12] By four months of age, infants are physically able to reach out and grasp an object, even if that object is partially hidden. But if, with the infant looking on, an object is totally hidden under a cloth or an opaque cup, the infant makes no attempt to get it back, acting instead as though the object no longer existed. Bower's surprising discovery was that infants behave in exactly the same way if the cup is transparent, the object in full view. For the infant, the cup's outer boundary determines the object; impenetrability prohibits there being anything inside it. What adults see as the inner object is apparently seen by the infant as a feature of the cup's surface. Infants, of course, do gradually learn to recover an object hidden under a cup, first if the cup is alone and then, by the end of the first year, if it is hidden under one of a small number of identical cups. Interestingly, however, infant performance on tasks like these is throughout the same for opaque as [it is] for transparent cups. The two play indistinguishable roles as the infant learns to retrieve hidden objects.

Other evidence for the infant's grasp of object impenetrability is more direct. In one series of experiments, sketched in figure 4, infants faced a wide shelf with a front section that could be folded back until it rested on the piece to which it was hinged. They were then habituated to the full range of motions of the shelf, as in 4a. Next a rectangular block was placed on the rigid rear section, and the infant was exposed to two test situations. In one, 4b, the shelf was rotated until it lay flat on the shelf, a trajectory that ought to have been prohibited with the block in place. In the other, 4c, the shelf was rotated just enough to have touched the block. From four and a half months, infants regularly stared longer at 4b, the display which

---

12. I follow the later summary provided in [T. G. R.] Bower, *Development in Infancy*, 2nd ed. (San Francisco: W. H. Freeman, 1982), chap. 7. This account differs significantly from that in the first edition of Bower's book.

Figure 4

violated the principle of impenetrability. A significant fraction of a group of three-and-a-half-month-old infants had already begun to do so.[13]

The preceding experiment shows that the infant expects the trajectory of a visible object, the shelf, to be constrained by the presence of an object that is invisible at the time the constraint is applied. Other experiments show that the same impenetrability constraint is expected even when neither object is visible at the time it applies. In one of these experiments, four-month-old infants were habituated to a display in which a ball was released above an occluding screen, fell behind it, and was seen, when the screen was raised, to be resting on the floor. That is the situation represented schematically in figure 5a, where the dotted line represents the screen. After habituation a false floor was introduced into the display a small distance above the real one, and the infants were tested on the two situations shown in 5b and 5c. In the first, the impossible situation, the ball was found under the false floor when the screen was raised; in the second, the ball was found resting on it. Infants looked for significantly longer at the anomalous situation in 5b. In a simplified version of the experiment, two-and-a-half-month-old infants behaved in the same way.[14]

These last experiments suggest a reformulation of the no-overlap principle that, though equivalent to impenetrability, was not suggested by it.

13. Renée Baillargeon, "Object Permanence in 3½- and 4½-Month-Old Infants," *Child Development* 23, no. 5 (1987): 655–64. Other experiments using a similar protocol show that, at least by seven months, infants have expectations that take account of the size, distance behind hinge, and compressibility of the occluded object. See Renée Baillargeon, "Young Infants' Reasoning about the Physical and Spatial Properties of a Hidden Object," *Cognitive Development* 2, no. 3 (1987): 179–200.

14. Elizabeth S. Spelke et al., "Origins of Knowledge," *Psychological Review* 99, no. 4 (1992): 605–32. For a more sophisticated version of the same behavior by somewhat older infants, see Renée Baillargeon, "Representing the Existence and Location of Hidden Objects: Object Permanence in 6- and 8-Month-Old Infants" *Cognition* 23, no. 1 (1986): 21–41.

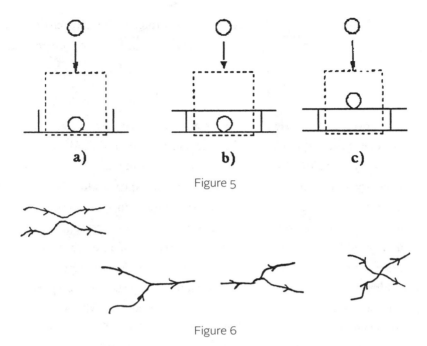

Figure 5

Figure 6

Conceived as a bounded region that moves through space over time, an object is a trajectory tracer, and traceable trajectories cannot have branch points. If we refer to an object's trajectory as its lifeline, extensible from birth or origin to death or dissolution, then the no-overlap principle becomes a prohibition of branch points or intersections of life lines, a prohibition illustrated in figure 6, where the directional signs indicate increasing time. The pair of lifelines shown in the upper left of the diagram are permissible, but the others are prohibited by their failure to satisfy the conditions required for objecthood. No two lifelines may ever occupy the same or overlapping regions of space. That is the characteristic which makes lifelines the uniquely suitable referents for proper names. Since lifelines may not intersect, attaching a proper name to a single position on a lifeline attaches it to the whole. Conversely, if two names (e.g., Tully and Cicero, Hesperus and Phosphorus) are attached to a single lifeline, then the objects that they name are one and the same. Slightly reformulated, the no-branch-point principle will prove applicable to kinds as well as to objects. It is then subject to failure, but only when kinds cease to behave as they should, the occasions that require the historian to turn interpreter. Though such occasions are relatively rare, they make kinds unsuitable to bear proper names.

II

Turn now from the basic-object concept to the accompanying concept of a kind. Both concepts are in evidence within hours of birth, and the latter is usefully introduced by noting what the former still lacks. The object as trajectory—or lifeline tracer—is bare of qualities: it ceases to exist when sensory contact with the track tracer is lost for longer than the few instants during which its motion can be extrapolated, e.g., as the object passes behind an occluding screen. The continued existence of an object that has disappeared from view is, however, intrinsic to the object concept as older children and adults know it. Indeed, what psychologists call *object permanence* seems a necessary characteristic of all objects, one without which they would not be objects at all. What the infant has still to learn is that objects are more than bounded regions of space and that their qualities— size, shape, color, texture, and so on—can ordinarily be used to reidentify them when they reappear after a moderate interval of invisibility.

Infants do recognize and make use of qualities, but until late in their first year that use plays no role in their object concept. Instead, the role of qualities is to permit the discrimination of situations calling for different behaviors. Among these discriminations, the earliest are elicited by the presence of important carepersons, most notably a child's mother. These early signs of recognition appear, however, to be evoked by recurrences of the mother-present situation, with no implication that that situation involves the return of an enduring object. Far from being the first object recognized by the infant, as one might suppose, mother is not, when first recognized by the infant, yet an object at all. Evidence for that conclusion is robust, but very recent. Before considering it, look briefly at reasons for believing that neonates recognize their mother's presence at all.

Though both smell and hearing sometimes play a role in the discrimination of the mother-present situation, visual reidentification by recognition of mother's face is a main tool from the start. Because more is known about it and because it will prove especially revealing, I here restrict attention to it. At the same early age (a few hours after birth) when human infants begin

to track moving objects, they also focus preferentially on a generalized facial pattern, two dots or blobs arranged symmetrically above a third blob.
The first two stand for eyes, the last for a nose or mouth, two features which
are visually merged in many animals.[15] Until the last decade it was widely
believed that a significant period of further development, both neural and
cognitive, was needed before an infant could learn to recognize an individual example of this pattern: a human infant, for example, was thought incapable of responding preferentially to its mother's face until an age around
two months. Recent evidence, however, suggests that that ability too is in
evidence a few hours after birth, at least in an immature form.

In typical experiments with infants between two days and five months
old, the child is placed in a supportive seat facing a curtained stage or display area in which, once everything is in place, the faces of mother and a
stranger are exhibited simultaneously through holes in the curtain. Even
infants as young as forty-eight hours, with as little as four hours of discontinuous exposure to their mothers, usually gaze significantly longer at her
than at the stranger. After establishing this initial preference for the child's
mother, one such experiment was extended by exposing the infant repeatedly to its mother's face alone. After habituation had been achieved, the
child was again exposed to mother and stranger simultaneously, and it
regularly gazed longer at the stranger.[16] In later months, the child gradually
develops the ability to recognize mother's face under less fully controlled
conditions (other angles of view, briefer exposure, dimmer lighting, and so
on), and its ability to recognize faces is extended to a growing number of
other significant individuals.[17] Further extension of its roster of recognizable faces occupies much of the child's later life.

15. Preference for the same pattern is observed in other animal species, including chicks.
See Johnson and Morton, *Biology and Cognitive Development*, 57–64, 104–6. This book is an
especially useful guide both to current views on face recognition and to the literature about it.

16. For the experiments on neonates see: I. W. R. Bushnell, F. Sai, and J. T. Mullin, "Neonatal Recognition of the Mother's Face," *British Journal of Developmental Psychology* 7, no. 1 (1989):
3–15; and Tiffany M. Field et al., "Mother-Stranger Face Discrimination by the Newborn," *Infant Behavior and Development* 7, no. 1 (1984): 19–25.

17. The development of face recognition after the first two days is not steady. In their chapter 6, Johnson and Morton (*Biology and Cognitive Development*) give reasons to suppose that
two quasi-independent reidentification systems are involved. One, predominantly subcortical, is present at birth, utilizes the peripheral visual field, and may well be associated with
the lifeline-tracing system described above. The other, which initially suppresses and then

It is difficult even to describe these experiments without implying that the infant is responding to the return of an object, and that is the way they have ordinarily been interpreted. But impressive recent evidence suggests that this is an overinterpretation. The most directly relevant was an unexpected by-product of experiments on infants' ability to imitate adult facial gestures.[18] That capability is in evidence as early as the first day after birth, and is well established by the sixth week. In the presently relevant experiment six-week-old infants were exposed in order to two adults, the child's mother and a previously unseen stranger. Whichever adult the child saw first would either stick out its tongue or else hold open its mouth, both facial gestures that such infants are able to imitate. The second adult would perform the other gesture. In previous experiments involving only a single gesturing adult, infants had regularly repeated the adult's gesture on successive exposure. In this new experiment, infants would often immediately repeat to the second gesturing adult the gesture made by the first. [They] behaved, that is, as though the first adult had simply reappeared. Only if the two adults appeared and disappeared on clearly different tracks would the infant wait for and then imitate the gesture of the second. It made no difference whether the first exposure was to mother or to the stranger, nor whether the initial gesture was mouth opening or tongue protrusion. Discontinuity of track rather than discontinuity of physiognomy was what signaled the arrival of a new and different individual. Though the infants could distinguish their mother from the stranger by face recognition, that discrimination did not imply change of identity, which was judged simply by continuity of track. It is as though mother were herself a kind, capable of numerous different exemplifications. Or, to revert to the viewpoint

---

supplants the first, requires the physical elaboration of the visual cortex during the early postnatal months. Daphne Maurer and Philip Salapatek, "Developmental Changes in the Scanning of Faces by Young Infants," *Child Development* 47, no. 2 (1976): 523–27, provides further information relevant to the distinction between the two putative systems. Since the distinction itself remains controversial, it should be emphasized that it plays no significant role in the present argument. Only the early establishment of a face-recognition system is important to the discussion that follows.

18. Andrew N. Meltzoff and M. Keith Moore, "Early Imitation within a Functional Framework: The Importance of Person Identity, Movement, and Development," *Infant Behavior and Development* 15, no. 4 (1992): 479–505.

introduced at the start of this section, it is as though mother-present were a frequently recurring situation.

Recent experiments by Fei Xu and Susan Carey with ten-month-old infants provide decisive evidence to the same effect, in this case for a larger variety of kinds.[19] In one experiment the infants were shown two objects of different kinds, the kinds being chosen from among those with which its parents were confident it was well acquainted (e.g., a cup, bottle, toy truck, wire basket). Two different objects were initially hidden behind a screen, and the infants were provided with a first exposure to them, one at a time. Next, the infants were habituated to the objects by a series of exposures, one object at one side of the stage, then the other at the other. Each object was left in place until the baby looked away for a significant interval, and the series was continued until baby's looking time was reduced by half. Then the baby was tested by removing the screen and observing the dependence of looking time on the number of objects, one or two, found behind it. The results were identical to those in a baseline experiment in which infants had shown some preference for two-object exposures. After habituation infants should have looked longer at the unexpected result, which for adults was exposure of a single object behind the screen. But habituation

19. Fei Xu and Susan Carey, "Infants' Metaphysics: The Case of Numerical Identity," draft dated April 28, 1994, of yet to be published paper [subsequently published in *Cognitive Psychology* 30, no. 2 (1996): 111–53]. Several other relevant experiments are described in this paper. Fei Xu[, Susan Carey,] and Jenny Welch, "Can 10-Month-Old Infants Use Object Kind Information on Object Segregation," draft for poster presentation, 1994 [subsequently published as "Infants' Ability to Use Object Kind Information for Object Individuation," *Cognition* 70, no. 2 (1999): 137–66.], get similar results with a different experimental design. Susan Carey, "Continuity and Discontinuity in Scientific Development," supplies additional information about children's ability to recognize and anticipate the number of objects in a display (late draft kindly supplied by the author). [Susan Carey (personal communication) does not recognize the last title that Kuhn quotes above, but she vividly remembers sharing her work with him, and his excitement about the connections that he saw between his own views concerning scientific development and the work of Carey and others on object individuation and emergence of "kind sortals" in infancy. The papers that she shared with Kuhn in draft form were all subsequently published. In addition to the paper cited above, see Fei Xu, Susan Carey, and Nina Quint, "The Emergence of Kind-Based Object Individuation in Infancy," *Cognitive Psychology* 49, no. 2 (2004): 155–90. See also Susan Carey's book *The Origin of Concepts,* Oxford Series in Cognitive Development (Oxford: Oxford University Press, 2009). I am grateful to Susan Carey for this information.—Ed.]

had no effect on the infants; neither one-object nor two-object displays were unexpected. They continued to behave as they had at the start.

In a second experiment the initial exposure was changed. In the infant's first exposure to the objects, prior to habituation, both were moved from behind the screen simultaneously, one being moved to one side of the stage, the other to the other. In contrast to the original procedure, above, the infants were at the start shown two simultaneous object tracks. Habituation and testing then went on as in the first design, but responses were very different. Two-thirds of the babies had preferred two-object displays before habituation; only one-third retained that preference after. Here again, though the ten-month-olds showed clear ability to discriminate the two kinds of objects in the tests, it was tracks, not objects, that they counted. As in the experiments on facial gestures, only object tracks, not object properties, were used to distinguish between objects. A variety of other experiments pointed in the same direction.

These results are particularly impressive when compared with what occurred when the first of the two immediately preceding experiments was performed with twelve- rather than ten-month-old infants. In the absence of any trajectory clues, these slightly older infants did show surprise at finding only a single object when the screen was raised. Twelve of sixteen twelve-month-olds looked longer at the unexpected (one-object) outcome, though only four of sixteen ten-month-old infants had done so.[20] In another sort of experiment twelve-month-old infants were repeatedly shown two different sorts of objects being placed in a box which hid them from view. Encouraged to retrieve them, infants reached twice into the box and withdrew both objects. Shown only one kind of object being placed repeatedly in the box, the infants reached in and retrieved an object only once. It appears that for twelve- but not for ten-month-olds, information about differences of kind can take the place in object counting of information previously supplied exclusively by tracking. Note also, for future ex-

---

20. This is experiment 5 from Xu and Carey, "Infants' Metaphysics." The experiment that immediately follows is reported in Fei Xu et al., "12-Month-Old Infants Have the Conceptual Resources to Support the Acquisition of Count Nouns" [in *The Proceedings of the Twenty-Sixth Annual Child Language Research Forum*, ed. Eve V. Clark, 231–38 (Stanford, CA: CSLI Publications, 1995).]

ploitation, that during the same two-month interval many of these infants had learned to recognize an increased number of the terms that named these various kinds.

III

I shall want shortly to ask about the nature of the conceptual changes to whose beginning these remarkable experiments attest. But it is essential first to consider the recognition process which they disclose. For that purpose I continue to concentrate on face recognition, which provides a particularly revealing introduction. Much enrichment and refinement aside, that process is the same in the first weeks of life, when what the infant recognizes is the recurrence of a member of the mother-present kind, as it [is at] the end of the first year, when the child recognizes the return of an enduring object, its mother. In either case, recognition results from the comparison of a new three-blob pattern with others stored in memory, and the nature of the comparison is also, in both cases, the same. This identity in processing is what makes possible the growing entanglement of individual and kind concepts underway at the end of the child's first year, an entanglement that quickly modifies the infant's object concept and, more gradually, its concept of kind as well. For the moment, I shall treat the two processes as one, speaking of both kind recognition and individual recognition as examples of identification.

Begin by noting that the term *recognition*, though it will be found to catch an essential characteristic of what has occurred, implies something unlikely. Furthermore, what is unlikely about it relates closely to what in chapter 3 made unlikely the standard view which describes meaning as a set of features characteristic of the thing meant. Though my present concern is with recognition and identification, the following remarks about them will anticipate the very similar remarks about meaning to be developed in the next chapter.

To say that Bobby, a child or adult, has learned to recognize the face of some particular individual *A* suggests a finished achievement: perhaps that Bobby has learned enough characteristics of *A*'s face so that, under good viewing conditions, he can pick out that face again from all others.

But what experiments tell us is not of quite that sort. Rather, they show Bobby's ability to *discriminate* a previously encountered individual A from some other individual X, again an achievement, but one far smaller than the standard formulation suggests. Bobby may be able to distinguish A from X but not from Y, or from X and Y but not from Z, or from X, Y, and Z but not from W, and so on. That series can be continued, and at each step along the way, Bobby will have learned one or more additional discriminations. But however many faces Bobby has learned to tell from A's in the past, he may still fail with the next individual he encounters. Learning to recognize an individual face is thus a process that can, in principle, continue indefinitely. Why should it have an end? No one needs to learn to discriminate more faces than he or she is ever likely to encounter. It will be time enough to learn additional discriminations when and if the need for them arises.

Viewed in this way, face recognition suggests with special clarity that neither identification of a particular individual nor of its kind requires knowledge of some special constellation of qualities, properties, or features shared by all presentations of that object or by all members of that kind. One need only know features that can differentiate the substance or kind in question from others with which it might, *in the world as it is*, be confounded. Recognition is not, as we shall see below, the only processes by which objects or kinds can be identified. But it is the basic one, and understanding that it proceeds from differentiae rather than shared features has a number of significant advantages. Within the restricted environment where they were acquired, the very limited store of differentiae in hand at early stages of the learning process are likely to produce correct identifications; the additional differentiae required for more general use may be acquired gradually with increasing experience of the world; the process of learning new differentiae need never come to [an] end. Additional differentiae, that is, make successful identification more likely, but they do not guarantee it; nor is their absence incompatible with it. There is nothing quite like having finally gotten the list of required features right. These advantages are, I suspect, all traceable to a single evolutionary source: differentiae, as a means of identification, are far less sensitive to error than characteristic features. Introduction of a feature that fails to differentiate reduces only the efficiency, not the accuracy, of the identification process;

introduction of a feature which subdivides kinds or kind members in ways not sanctioned in the user's community will, as I shall later argue, be rapidly identified and corrected during the process of language learning.

Very little is known about the set of features actually used in identification by either humans or animals. Probably there are many sets that can serve the same function, and no two individuals need use the same one. But some things can be said about them, and face recognition will again suggest what they are. The only shared feature prerequisite to recognition of a particular face is the characteristic three-blob pattern, and that feature is shared by all faces. Other characteristics are relevant only to the extent that they provide indices of differences between the face to be identified and others. Among the most obvious of these are hair color, eye color, and skin color, but none need be known. (I can, myself, never remember the eye color of even the people to whom I am closest, and I have great difficulty remembering the hair color even of people I know quite well.) Far more important are the differentiae which caricaturist[s] exaggerate to produce schematic sketches that are unmistakable likenesses of their subject. Unlike hair and eye color, few of these differentiae have names, and, in the absence of a caricature, few people are able to describe them.[21]

Plausible and frequently suggested differentiae for face recognition share important features with the differentiae exploited by caricaturists. The standard candidates are usually specified as ratios: face height to face width, eye separation to face width, eye separation to distance between eyes and mouth, and so on. Apparently the brain can recognize and compute such ratios without the mind's being able to identify what they are. They play no role in verbal descriptions except when they are aberrant in the extreme. Unless professionally engaged in measuring them, no one can specify the

---

21. Terry Landau, *About Faces* (New York: Anchor Books, 1989), 45–48, describes and shows the results of a computer program that generates caricatures by comparing a drawing of a normal face with a drawing of the face to be caricatured. That approach emphasizes the essential role of the differentiae generated by comparison between faces. In the ordinary learning process, however, all that need be compared are actual faces drawn from the community in which reidentification is to be carried out. The "normal" face (of members of that community) is a product of many such comparisons, not something that the learner requires in advance. Reliance on a normal face, like reliance on prototypes more generally, still too much resembles the traditional shared-feature approaches to problems of identification and meaning.

value exhibited by an individual, even for those he or she knows best. And, finally, most or all of them are learned and refined only in the course of learning to recognize faces. Not surprisingly, then, the differentiae actually acquired through practice depend on the faces to which the learner is exposed. To say that a person adept in distinguishing faces within one culture, tribe, or race may initially find that members of another all look alike has become a cliché.

I have been urging the advantages of conceiving recognition as a process achieved with differential rather than shared or characteristic features. But the latter, identification by shared features, owes its traditional status to an apparently overwhelming advantage. If recognition proceeds from a list of necessary and sufficient conditions or something else of the sort, identification becomes routine: one simply checks the features of the object to be classified or identified against the list to determine whether the corresponding identification fits. No procedure of anything like this sort can help with identification by differential features. What method is used instead? The answer to that question will suggest what makes *recognition* the appropriate word for the process.

Think again of the process of recognizing a face, say mother's. The differentiae with which it is carried out are chosen to maximize the perceived differences between her face and the faces of others previously encountered. If the features being employed are thought of as the dimensions of a space (something that is especially easy to do with ratios like those presumably used in face recognition), then the positions of the faces of different individuals will be maximally separated, while the positions of different appearances of the same face will lie close together. In that space a new presentation of mother's face will lie in or near the cluster of previous presentations and at a distance from the presentations of the faces of others. Mother can then be picked out at a glance, not by virtue of any particular feature she possesses, but because, in the space of differentiae being employed, there is no one sufficiently like her to cause confusion. Nothing like an inference to the best or most plausible hypothesis is required, for in these circumstances no alternative hypothesis is in question. One cannot imagine being mistaken, and, if one finds that one has been, one is shocked, as though the world had betrayed one. "How could I," the stunned observer may ask, "conceivably have mistaken that hag (or that gorgeous chick) for my mother?" Some readers may recognize that shock as a relative of the ones illustrated in *Structure* by talk of gestalt experiments and of the world's changing. We shall return repeatedly to the examination of such experiences in later parts of this book.

Recognition by differentiae is not, of course, the only process through which identification of kind or reidentification of substance can occur. There are many circumstances (e.g., dim light, distance from the object of identification, mislaid eyeglasses) in which one needs to ponder what one has seen, examining the particular features of its presentation ("Oh yes, that's the dress that mother wore this morning"), and infer the relative likelihoods of various identifications. But recognition is the more primitive process, and it remains basic when inferential processes are developed. Of the two, it is the faster, surer identification process, and it in any case supplies the domain of recognizable objects and situations that inferential identification processes require. Much of this book will be required to justify that nonstandard claim, but chapter 3 anticipates the nature of the required justification. The domain over which inference can occur is limited by what I there called a *kind set*. Both the probability and the truth/falsity of any particular inference are thus relativized to a kind set, and any individual kind set makes it impossible even to conceive some parts of the domain made available by another. There can be no inferences whose domain encompasses all the worlds that may ever be conceived. Both the integrity of the recognition process and its status as a prerequisite to inference are thus basic to and also outcomes of the arguments of this book. In that sense the argument for them is circular, but the circularity will not prove vicious.

This account of recognition does, however, depend critically on the assertion that, in a space of suitably chosen differentiae, different presentations of the same individual or [the same] kind form a cluster at a distance from those formed by presentations of other individuals or other kinds. That statement amounts to a qualitative no-overlap principle, and for kinds it is just the sort of principle promised in chapter 3 as key to the explanation of incommensurability. In its present form it applies only to the kinds available before the acquisition of a developed language: the elaboration it requires to function with the far more complex arsenal of kinds acquired with language will be considered in chapter 5. Once again, however, the prelinguistic form is basic, and it will help to ask both about the evidence for its existence and about its likely source.

As to its source, the presumptive answer is implicit in what has already been said. Primitive kinds function primarily or entirely to discriminate situations requiring different behavioral responses. Differentiae supply the fastest and surest means to that end, and the apparatus within which they function is likely to be a very old product of evolutionary development, subcortical and present throughout the animal kingdom. The kinds which that apparatus discriminates and the responses which it matches to them vary, of course, from species to species and environment to environment. Some of them may be present at birth but others are learned from adults who already know them.[22] That there is some such situation-discriminating apparatus should cause no surprise.

But what evidence, other than evolutionary plausibility, supports the claim that that apparatus functions by differentiae? What evidence is there that the basic form of identification functions by clustering presentations of members of the same kind at a distance from the clustered presentations of members of other kinds? For me the strongest evidence simply reverses the direction of the route which has led to the claim. Conceiving kinds as identified by differentiae opens the way to a long-sought theory of the meaning of kind terms, and that theory brings with it a way of explicating the experience of incommensurability and of describing the changes that I, thirty years ago, described as scientific revolutions. But there is also evidence of a more direct sort.

That evidence is drawn from an area of psychological research known as categorical perception.[23] First recognized in phonological studies a quar-

22. For prelinguistic correction of errors in the application of kind concepts, see Cheney and Seyfarth, *How Monkeys See the World*, 129–37.

23. A very full and almost up-to-date survey of work in this field is provided by Steven Harnad, ed., *Categorical Perception: The Groundwork of Cognition* (Cambridge: Cambridge University Press, 1987), which includes a series of very full bibliographies. For what follows, its chapters 3, 4, and 5 have proved particularly useful. In order, they are Bruno H. Repp and Alvin M. Liberman, "Phonetic Category Boundaries Are Flexible"; Stuart Rosen and Peter Howell, "Auditory, Articulatory, and Learning Explanations in Speech"; and Peter D. Eimas, Joanne L. Miller, and Peter W. Jusczyk, "On Infant Speech Perception and the Acquisition of Language."

ter century ago, it has since been discovered also in a number of other areas, especially [in] the perception of music, color, and perhaps also facial expressions. The field is extremely active and still characterized by much controversy. But the status of its core findings [is] unchallenged, especially in the perception of speech, the area that has been most studied, and to which I largely restrict myself. The simplest form of categoric perception occurs when subjects exposed to a stimulus that can be varied over a wide range divide that range perceptually into two or more subranges within each of which perceptions are very similar, though they differ markedly from those experienced in a neighboring subrange. Color perception will provide a sense of what is involved. Normal trichromats can perceive colors stimulated by monochromatic light over the entire range between the infrared and the ultraviolet. Throughout that range, they can identify their perceptions with considerable unanimity using just the four color terms *red*, *yellow*, *green*, and *blue*, either alone or in two-word combinations. Furthermore, if asked to say whether two colors separated by a small wavelength interval were the same or different, their perception of difference is far more acute in the region between colors, say yellow and green, than in the single-color ranges on either side of that discriminatory peak.[24] These are the two characteristics that lead to the title *categorical perception*: a continuous range of stimuli is perceived as subdivided into discrete subranges, and the ability to discriminate small stimulus difference is markedly greater at the boundaries between subranges than within the subranges themselves.

A less familiar example, this one drawn from speech perception, will clarify and extend these points. Spoken language makes much use of stop consonants or plosives, sounds the utterance of which begins with the sudden opening of the air channels from the lungs to the environment. They come in three pairs corresponding to the location of the initial blockage.

Other bits of information referred to below are scattered, but they can readily be retrieved through the volume's excellent index.

24. Marc H. Bornstein, "Perceptual Categories in Vision and Audition," in Harnad, *Categorical Perception*, chap. 9; see also [Steven Harnad, "Category Induction and Representation," in the same volume,] 535. To the best of my knowledge, it has not been established whether similar differences in the ability to discriminate exist also at boundaries like that between brown and grey. If they do, they must vary with culture, for not all cultures make use of these two color categories.

For /p/ and /b/ the blockage is caused by closed lips; for /t/ and /d/ by the tongue; and for /k/ and /g/ by the glottis. Within each pair, the members are distinguished by the rapidity with which the release of blockage is followed by a vowel sound in their enunciation. In both *path* and *bath*, for example, the initial consonant is followed by the same vowel, but the initial consonant in *path* can be enunciated without a following vowel while in *bath* the omission of the vowel transforms the /b/ to a /p/. The first member of each of the preceding pairs is therefore referred to as voiced, the second as unvoiced. In use, both unvoiced and voiced consonant are ultimately followed by a vowel, but the lag—known as the voice-onset time or VOT—between plosion and voicing is longer for the first. All six stop consonants can be artificially synthesized, the VOT varied in small steps, and listeners asked either to identify the sound they hear or else whether they can distinguish sounds with specifically different VOTs. Their responses provide a striking demonstration of categorical perception.

For simplicity I restrict attention to /p/ vs. /b/ discriminations. If the lag time for voicing starts long, say around 90 ms, and is then decreased 20 ms at a time, listeners initially report /p/ as the initial consonant, and they discriminate examples with lags 20 ms apart only about 25 percent of the time. At a lag of around 30 ms, however, the initial consonant is perceived as switching rapidly to /b/, and it continues to be heard that way as the lag is further reduced, soon taking on negative values. If neighboring stimuli lie on opposite side[s] of this discriminatory peak, subjects are able to discriminate them about 70 percent of the time. These findings and many others like them display both what I have been calling *recognition by discriminatae* and also the clustering on which the recognition process depends. Stimuli on opposite sides of the discriminatory peak are heard as distinctly different, but those lying on the same side of the peak are heard as very nearly the same.

As to the location of the boundary, experiments both with nonspeech sounds and with young infants suggest that, though humans enter the world with at least some of them in place, experience with the speech in their environment can move, eliminate, or replace discriminatory peaks already present. In adult communities, therefore, boundary location depends markedly on language. The observed peak for English at 30 ms lag is apparently an enhancement of one present from birth. But for Spanish the peak occurs around 10 ms, and for Thai there are two peaks. Differences that are unmistakable to speaker[s] of one of these languages are often unheard by speakers of another. (Think of the problems that Japanese speakers have with the distinction between English /r/ and /l/.) Finally, though lag in the onset of vocalization is the primary determinate of the boundary between the paired consonants, many other variables—e.g., the location of the initial stop, a preceding fricative noise, and so on—also play a role in its location. It is largely for simplicity that I have here presented the space of relevant differentiae as one-dimensional.

In these and other cases of categorical perception, the point at which I have been aiming is inescapable. One does not *infer* a color one has seen or a sound one has heard. (What would [one] infer either of them from?) These are paradigm cases of *recognition*, and I am suggesting that both kinds and objects are also often picked out by recognition: to see them is to know them.

IV

Return now to the change in infant behavior between the ages, roughly, of ten and twelve months. What shall we say has happened? Available evidence permits no sure answer, but the account that follows fits what is currently known and suggests directions for future exploration. We have seen that nongeometric properties play no part in the infant's object concept during its first eight months or a bit more. What are objects for adults are for [the infant] simply bounded regions of space, in motion with respect to their background. Even the size and shape of the moving region play no part in determining object identity, much less such qualities as color and texture. When out of sensory range, these proto-objects cease to exist for the infant, or that is how adults describe its behavior. One may suspect, however, that that description misses a central point: What, in a world of tracks and situations, would existence be?

[T. G. R.] Bower, for example, believes that infants first attain the concept of object permanence around the age of six months, because they will then retrieve an object that they have seen hidden under a cloth.[25] But he adds that the infant in this stage "still seems to have a peculiar concept of

25. Bower, *Development in Infancy*, 195–205.

objects": it will, he points out, still look for an object in the place it was originally hidden even after seeing it removed and rehidden in another place.[26] These experiments, however, like a number of those previously discussed, are better understood by supposing that the child who removes the cloth from the place where the body was originally hidden is trying to re-create the body-present situation that existed before the cloth was put in place, much as it recreates the mother-present situation by screaming. If an adult has not moved the object meanwhile, the child is successful; otherwise it fails. But even when successful the child need not conceive the object as the one originally hidden. Indeed, it does not appear that its object concept has a place for same vs. different. Even when it extrapolates motion on a track during a brief period of invisibility, it may recognize same track without concluding to same object. Though there is no better term, describing what the child tracks, calling it an *object* has proved badly misleading.

Among the changes that occur, or begin to occur, between the ages of ten and twelve months, the most obvious is the application to proto-objects, previously known only by their tracks, of the recognition mechanism previously reserved for kinds. That change makes it possible for the infant to reidentify particular track makers by their qualities after sensory contact has been reestablished. The child's behavior then no longer suggests that the object has gone out of existence when it disappears; object permanence has been established. The process of change is not, however, merely the empirical discovery of an additional property of proto-objects. The change from proto-object to object is only one of a series of interrelated conceptual changes that the child undergoes at this time, and these changes together cannot be simply empirical in the sense of being learned by direct exploration of the world. Doubtless they are due in part to the development of additional cortical apparatus during the child's first year. But they are almost certainly also associated, perhaps through that new cortical apparatus, with the early stages of language acquisition. Before sketching the other conceptual changes which accompany object permanence, let me supply reasons for supposing that language plays a central role in the shift.

26. [Bower, 198.]

The experiments of Xu and Carey, previously discussed, provide the only direct evidence of which I'm aware. For ten- and twelve-month-old infants, there is a strong correlation between their knowledge of kind terms (e.g., *ball, bottle, cup, book*) and their ability to individuate kind members, not just by their tracks, but by their qualities. One plausible explanation is easy to find, and there are doubtless others of the same sort. A child who knows, say, the word *cup* soon learns that it has several referents, often differentiated by their qualities. The same properties that differentiate them also make it possible to recognize them, and that reidentification by qualities brings object permanence with it. As this occurs, individual cups become objects in the full sense of the term, and kinds cease to be exclusively kinds of situation and are extended also to kinds of object.

Though additional direct evidence for the association of language acquisition with object permanence is badly needed, evolutionary considerations supply persuasive evidence of another sort. For purposes of life without language, there is no apparent need to transcend the conceptual vocabulary of tracks and situations. The ability to recognize situation kinds, including the presence of significant individuals like mother, permits the tuning of behavior to the presence of friend and foe, to the various kinds of predators, and to a variety of quarry. Also, it permits the emission of a characteristic cry that alerts conspecifics within range to the situation that has arisen, thus increasing the species' ability to survive. Given this competence in recognition, the tracking response can match behavioral responses—e.g., approach and avoidance, pursuit and flight—to situations. In the absence of language, no useful function would be served by the ability to identify not just a particular kind of quarry or predator but a particular member of that kind. Language, on the other hand, seems to require that capability. In the absence of object permanence, language would lack what Ruth Millikan has called a proper function.[27]

An essential function of language is to enlarge the subject of possible communication beyond the physically and temporally present. Pointing and other gestures, together with coded cries, are well suited for communication

27. Ruth Garrett Millikan, *Language, Thought, and Other Biological Categories: New Foundations for Realism* (Cambridge, MA: MIT Press, 1984). This important book is among the few works that suggests what philosophers might gain by taking evolution seriously. I cannot specify my debt to it, but it's likely to be large.

about present tracks and situations, but they can go little further. When mother becomes an object, however, not just a recurrent situation, then she may, for example, be in the kitchen or bedroom. Such information is useful, but it implies object permanence, and language is required to communicate it. No wonder the two develop together. Object permanence cannot, however, develop alone. It requires, for example, a distinction between [being] out [of] sight and [being] out of existence. And that distinction is still only a start. Like "in the marketplace," the phrase "in the kitchen" supplies a location or place, an answer to the question "Where?" That question cannot sensibly be asked before the concepts of object, place, and time—the three Kantian components inextricably bound together in the neonate's tracking response—are disentangled, a disentanglement that also appears to come with language. I do not suggest that all these changes occur during the interval between ten and twelve months of age. Indeed, their full development must occupy many months, possibly more than a year of uncertain exploration. But I do suggest that, together with object permanence, they are the intertwined elements of a package that cannot, if conceptual stability is to be preserved, be unpacked one piece at a time.

One aspect of the content of that package is of particular importance to the concerns of this book. Kinds, as they entered this chapter, were always situation kinds, members of which were identified by their qualities. They might better, to avoid misleading, have been called simply *situations*. The basic technique by which they were identified was recognition, a noninferential process made possible by locating kind members in an appropriate field of differentiae where members of the same kind clustered at a distance from the clusters formed by members of other kinds. Use of that technique subjected kinds to a sort of qualitative no-overlap principle: no particular situation could be in two clusters, belong to two overlapping kinds. Unlike the geometric no-overlap principle, the qualitative principle was normic rather than nomic. It could, that is, admit exceptions, though only at a cost to which this book will repeatedly return.[a] When, at the end of the first year, that same recognition technique is applied to the reidentification of objects, it brings the qualitative no-overlap principle with it, and in its new application the qualitative principle is entangled with the geometric, nomic principle that prohibits the intersection of lifelines. The result is a simultaneous increase in the force of the normic principle and a quasi-logical difficulty in conceiving the division of any species of lifeline tracers. How can one species become two without a similar (and prohibited) division of its individual members? That question has answers, to some of which we shall return, but they were not easy to find.

V

By now it must be apparent that this chapter has, among other things, attempted an extended example of the interpretive process illustrated in chapter 2, the process required to set the stage for the beginning of any developmental narrative. There is, however, a noteworthy difference between this example and the three that preceded [it]. In all four cases, the objects to be interpreted were bits of behavior. But in chapter 3 that behavior was both expressed and interpreted in language, while in this chapter it is prelinguistic and may not be interpretable in language at all. That users of one language can with adequate effort always learn to speak and understand the language used by another group of users is probably guaranteed by the shared biological heritage of language users and by the primary function, descriptive communication, which language serves for all of them. Though that guarantee does not assure full translatability, a point to be further developed in later chapters, it does promise bilinguality, a limited tool for communication but a powerful one for understanding. One cannot, however, become bilingual in a nonexistent language, and the lack of that access may greatly limit the interpretations that one's own language can provide. I close this chapter with two examples. The first is drawn from this chapter and illustrates the difficulties that creatures endowed with language (ourselves) confront in understanding the behavior of those without it. The second reverses the direction and suggests how the neural apparatus developed for life before language constrains what can coherently be put into words in any language at all.

Early in this chapter I described the tracking response as evidence for the existence of the basic-object concept, later to declare that until track makers could be reidentified by their qualities, one should not speak of an "object concept" at all. Elsewhere, I questioned the appropriateness of distinguishing between disappearance and going out of existence when interpreting the child's behavior toward what we adults call objects. To speak of tracks and track makers from the start might have obscured the difficulty, but it would in no way have reduced it. Tracks are nothing if they are not routes traced through space over time, and it is not clear that anything but an object could trace them. It is these difficulties in interpretation that led me to suggest that the tracking response in infants and in animals implies no cognitive separation of what, for language users, are the concepts of space, time, and object. That suggestion may prove mistaken, but the only presently available argument against it is the inability of linguistically endowed creatures to imagine what life in the absence of that separation would be like. That is no argument, however, only a higher form of ethnocentrism.

Compare the examples of chapter 2 with the one developed in this chapter. In all three of the earlier examples, I began by using familiar terms to describe the beliefs of the author of an old text, whether that author was Aristotle, or Volta, or Planck. Then I insisted that those terms, used in that way, made many passages of the texts in question anomalous. Other meanings, which I attempted to supply for readers, would remove those anomalies, giving evidence that meanings had changed since the texts were written. In this chapter I began in the same way, using familiar words in familiar ways and then pointing to the behaviors that made those words anomalous. But what I have not done and cannot even imagine doing is suggest alternate meanings that would remove the anomalies, making the behavior conceptually comprehensible. Doubtless the particular form these difficulties have taken for me result partly from a shortage of evidence and partly from my own inexperience in handling what evidence there is. My account of these prelinguistic developments will surely be both extended and corrected. But I think it unlikely that the aspects of my discussion most relevant to this book will be markedly altered. Where there is no language, the possibility of the sorts of interpretation that produces insight and understanding is extremely limited.

That thesis has a significant converse: resort to language limits what can be understood. The next chapter will show how these limits operate between users of different languages, introducing the concept of incommensurability for that purpose. Much more will first be said about kinds in order to prepare the way. But the discussion here of the prerequisites to language permits an example of the limits imposed by use of any language at all. All languages, I have been suggesting, have at their foundation neural structures evolved to permit survival in a world in which distinctions made possible by language had no function. As such previously functionless distinctions evolved—some with one language, others with another—most could simply be added to the older distinctions embedded in the preexistent neural base that all languages share. But others, though discovered by the use of language, could not coherently be embodied in it, and the difficulties encountered when trying to talk about them are the same as those encountered in this chapter when trying to talk, say, about a prelinguistic object concept. The so-called paradoxes of the quantum theory provide an apposite example.

Lifeline tracing is basic to the procedures we have been examining, nonlinguistic as well as linguistic. An immense amount of accumulated human experience testifies to the extent of its utility and use. Nevertheless, there is now excellent reason to suppose that the procedures which are based upon it cannot be adapted to the microworld. Electrons, protons, and other subatomic particles cannot be individuated by lifelines; they are not simply very small versions of the physical bodies of the everyday world. In an important sense, they are not particles at all, but that should not suggest that they are something else instead. All attempts to describe what they are have required resort to intrinsically incoherent locutions of which *wavicle*—partly like a wave and partly like a particle—is the best known. Others arise in explications of the Heisenberg uncertainty principle. For years I have expected that such incoherencies would vanish with the invention of concepts and language more appropriate to the subject matter. But I now think nothing of the sort can occur.[28] Though the particles of the microworld are as real as things get, and though we can tailor our interactions with them in great detail, neither our language nor our conceptual apparatus [is] likely ever to have a comfortable place for them.[29]

February 28, 1995

---

28. This is a viewpoint first presented by Niels Bohr. [See Bohr's *Causality and Complementarity*, vol. 4 of *The Philosophical Writings of Niels Bohr*, ed. Jan Faye and Henry J. Folse (Woodbridge, CT: Ox Bow Press, 1998).]

29. Ian Hacking, *Representing and Intervening: Introductory Topics in the Philosophy of Natural Science* (Cambridge: Cambridge University Press, 1983), provides a splendid account of the reality of microparticles by describing the nature of human interactions with them.

CHAPTER 5

Natural Kinds

How Their Names Mean

By the end of the first year of life human infants have begun to acquire language and also, in the process, to reorganize their concepts both of objects and of kinds. In that process objects come to be understood as track makers which come and go, their identity being redetermined by their qualities when sensory contact with them is interrupted. Infants at this early stage may recognize no more than two distinct kinds: objects, on the one hand, and the situations to which they adjust their responses, on the other, and little is known about the ways in which further development occurs except that *interactions* with adult language users [remain] central to most or all of it. But no presently relevant function would be served by continuing the developmental narrative required to recover the evolutionary foundation of kind/object concepts. Instead, I shall now assume an adult's acquaintance with the kinds and kind terms most relevant for scientific development, and ask about the main elaborations needed to make the year-old child's concept of kinds conform to it. In the process, I shall be developing the notion of a structured kind set, the concept announced in chapter 3 as required for understanding the nature of the special sorts of belief change illustrated by example in chapter 2. Scattered glimpses of the altered understanding of scientific knowledge that results will appear along the way.

Adult speakers, of course, recognize a variety of kinds far larger and more varied than can be dealt with here, and I shall therefore restrict attention to the kinds most relevant to the development of science. Three of them, all manifest in everyday discourse, play an especially fundamental role. First, and probably basic to the emergence of the other two, are the living organisms of everyday life: people and animals, trees and plants. All make tracks against a background, either because they themselves move (people and animals) or because the occluded background changes with the motion of the observer (trees and plants). In addition, all endure for extended periods of time during parts of which all sensory contact with them may be lost. The second major variety of kinds is the materials from which objects are made: wood, rock, flesh, bone, gold, water, etc. Like animals and plants, materials endure through time, but they are not objects and they do not make tracks. While kinds of animals and plants are named by count nouns, the names of materials are mass nouns. Despite that major difference, these two sorts of kinds share many essential features. Most or all of the so-called natural kinds belong to one or the other of these two, which will both be discussed in this chapter. The third variety of kinds are everyday artefacts: cups and bowls, tables and chairs, shoes and gloves, bats and balls, bicycles and automobiles, knives and screwdrivers. Like living creatures, they are track makers and named by count nouns, but their reidentification and classification require techniques very different from those used for natural kinds. Their discussion is reserved for the next chapter.

Even this much-restricted variety of kinds suggests the complexities that lie ahead. Kinds of material are not, for example, objects or track makers, but they do seem to obey the no-overlap principle in the simple form previously stated: no material can be both wood and water, or so I shall later argue. But for objects the no-overlap principle is more complex. A frying pan can, for example, be both a utensil and an iron object; a dog can be an animal, a male, and a pet. Considerations like these take us well beyond the consideration of objects and situations in chapter 4. Nevertheless, all kind concepts appear to have been shaped by the object concept, and several of the themes developed in that chapter will continue to prove basic. Reminders about three of them should be a helpful prelude to the development of a theory of kinds, one primarily restricted for the moment to so-called natural kinds.

First, objects are track makers, and there are two necessary restrictions on the shape of their tracks: the track of a single object may not have branch points; the tracks of distinct objects may not intersect. Second, objects must be reidentifiable by their qualities when they reappear after an interval long enough to bar reidentification by continuity of track. A track maker which cannot in principle be reidentified in this way is not an object at all. Some such principled distinction between the reidentification of a previously seen object and the identification of an indistinguishable new one is presupposed by many of the practices of everyday life: for example, the operations of arithmetic. Only if the members of a collection remained simultaneously in view could one distinguish between an enumeration of its members (e.g., the number of fishes in a tank) and a count which included some members repeatedly. Note, however, that this guarantee breaks down in quantum mechanics together with the object concept itself. Certain kinds of fundamental "particles" are in principle indistinguishable, require new ways of "counting," and display correspondingly novel statistical regularities.

The third consequential theme introduced in chapter 4 was more tentative. Though the ability to reidentify an object must be learned, it does not require acquisition of anything so strong as a set of features characteristic of that object and of it alone. Encountering a putative new object, a person need learn only to distinguish it from the other objects he or she has previously learned to reidentify. Reidentification may occur, that is, in a space of differentiae which is steadily enriched as the user encounters (and learns to reidentify) new objects. In the discussion of objects, however, the evidence favoring differential over characteristic features was relatively limited. Introduced primarily in connection with face recognition, it took two forms. First, repeated attempts have failed to suggest what characteristic features would be required to single out one particular face (or object) in a universe of all possible faces: to specify, for example, the necessary and sufficient conditions for that face's being, say, Susan's, or for that animal being my cat. And, second, the usual emphasis on necessary and sufficient conditions makes the reidentification process gratuitously elaborate. Learning to reidentify a particular face or a particular object is more plausibly regarded as a continuing process requiring a space of differentiae that must, for some time, be steadily enriched as the learner encounters more and more objects that must be told apart.

These same arguments apply to the identification of kinds of objects, but in the case of kinds a great many other arguments are available, most of them resulting from the puzzles, recognized and unrecognized, that are resolved or dissolved in the course of developing a theory about them. That task is the subject of this chapter, and I shall approach it by assuming from the start that kind membership is established by differentiae rather than characteristic features. Much further evidence for that assumption will accumulate as we proceed.

I

As a first step toward a more general theory of kinds, I begin with living organisms of a size visible to the naked eye. Like nonliving track makers, these creatures owe their status as objects to the fact that they can be reidentified after a period of sensory invisibility. That is true also of nonliving objects, and much of the argument that follows applies to them as well.[1] But the reidentification of living creatures presents a special problem: the qualities relevant to their reidentification change as individuals move along their lifelines from birth to death. In practice, that change is often sufficient to bar reidentification, but objecthood does not require reidentifiability over intervals which include large-scale alterations. Qualitative changes occur continuously and usually slowly; objecthood requires only piecemeal reidentification as qualities change. For any object, that is, there must be some finite interval Δt over which reidentification is always possible wherever on the object's lifeline that interval is placed. If that condition is met, permanence along the whole of a lifeline can in principle be ensured by a

---

1. The names of kinds of objects are, in philosophy, often referred to as *sortals* or *sortal predicates*, and I am deeply indebted to literature on that subject for much in this chapter and for bits in the next. Two particularly helpful books are David Wiggins, *Sameness and Substance* (Cambridge, MA: Harvard University Press, 1980), and Eli Hirsch, *The Concept of Identity* (Oxford: Oxford University Press, 1982). Much of my emphasis both on objects and on their reidentification dates from my encounter with the first of them (cf. chap. 4, n. 10). The emphasis on continuity of changes comes particularly from the second. That difference in emphasis between them is best understood as resulting from the tendency of the sortal literature to treat objects as all of a piece, letting the choice of favorite example (donkey for Wiggins, automobile for Hirsch) direct the discussion.

succession of reidentifications over the stipulated short interval. An object for which no such interval existed would not, ipso facto, be an object at all.

The necessity of a reidentifiability interval Δt for objects restricts the variety of objects that can exist in the world, but the strength of the required restriction depends on the manner in which reidentification is achieved.[2] If, as has usually been supposed, objects could be reidentified by virtue of some set of features that they and they alone possess during Δt, then the existence during that same interval of an indistinguishable object must be prohibited.[3] But if, as I am presently supposing, reidentification is achieved by differentiae, a much stronger condition, equivalent to Leibniz's principle of the identity of indiscernibles, is required. The appearances of the object, as observed at the two ends of the interval Δt, must in this case be compared, not simply with each other but with the appearances of all other objects present at the time or remembered from some previous first- or secondhand encounter. That comparison must show the initial pair to be more like each other (to cluster more closely) than either is like (or clusters with) the appearance of any other known object. If there had at any time been an object able to leave a memory trace indistinguishable from that left by the candidate for reidentification, then it would be impossible for someone who had encountered both to tell which had disappeared or reappeared. If such an indistinguishable object came into being in the future, then the older object would create the same uncertainty for reidentification of the new one. Objecthood thus requires that no two indistinguishable objects exist anywhere in the universe, unless separated

2. Knowledgeable readers will ask at once whether the restriction is metaphysical (a restriction on the variety of objects that can exist) or epistemological (a restriction on the variety of objects that can be known). I want to insist that, because the restriction is required for the viability of the object concept, its metaphysical and epistemological aspects are inseparable. We cannot know that something is an object unless there is a corresponding Δt, but also, in the absence of some Δt, that something is not an object. On this point the paradoxes of the quantum theory are again instructive.

3. This statement is in the subjunctive because the apparently insuperable difficulties in accounting for reidentification in terms of characteristic features apply equally across the short interval, Δt, and the longer intervals across which reidentification is generally called for. I take the former process to be impossible in principle, but it has long supplied the standard way of thinking about reidentification, and the depth and centrality of the conceptual changes which result from the transition to differentiae require special emphasis.

by a time interval so large that their lifetimes could not possibly be simultaneously present within it.[4] Two indistinguishable objects are either the same object, or neither is an object at all.

The reidentification of objects requires, then, a space of qualitative differentiae in which the candidate for reidentification lies closer to—clusters more closely with—some object that has disappeared than with any other object in memory or in present view. The dimensions of that space might, in principle, be chosen so that each and every individual could be distinguished from all others. But its dimensionality would then be vast, and the search of memory required to locate the cluster in which to place a previously seen but freshly presented track maker would be prohibitively long. It proves possible, however, to divide track makers into kinds within each of which reidentification can be accomplished using a far more limited number of differentiae, the appropriate set and the applicable Δt varying from one kind to another. Indeed, as the closing pages of the last chapter suggest, the ability to reidentify individual objects and to discriminate kinds of objects are closely related and appear to emerge together in the last months of the first year of life. Both are applications of the ability to cluster appearances in a field of differentiae. In the first, different appearances of the same individual are clustered at a distance from the clustered appearances of other individuals. In the second, appearances of different individuals of the same kind are clustered at a distance from the clustered appearances of individuals of other kinds. Development of these two applications of that ability appear to build on each other during the development of the individual who makes use of them.

Figure 1a will begin to suggest what is involved. It shows a flock of ducks displayed in a space of differentiae chosen so that they can be distinguished from each other. That distinguishability is indicated by the white space separating them, and that white space is required by what I have called

4. This condition, like the one given for reidentification by characteristic features, can be given a somewhat weaker formulation if the relativistic limits on possible communication are taken into account. The argument becomes much more cumbersome, however, and I think no new points of principle are introduced. More will be said below about the restriction to nonoverlapping lifetimes.

Leibniz's principle, itself a precondition for their objecthood.[5] Note that the dimensions of this space need not include the features we all expect of ducks—feathers, for example, or webbed feet—for these are shared by all ducks and by other creatures beside[s]. What is required [are] features in which individual ducks differ, and, as was the case for face discrimination in chapter 4, little systematic or general is known about them. But I am fortunate in a friend who took care of the family's ducks in her youth and who tells me she found especially useful such features as marking around the eyes, curl of the "smile line" that runs from the base of the beak onto the sides of the face, general distribution of weight (broad-beamed, for example, or top-heavy), and length of legs for a given body size. Doubtless many other features can play a role in duck reidentification, and there is no reason to suppose that all who can call ducks by name use the same ones. It says something about the process, however, that in my friend's case, color, though useful for distinguishing age, played little role in reidentification: her ducks were all of one breed.

So far, we have been concerned with the reidentification of objects, but that concern has now brought us to the point that motivated [the] introduction [of this topic], [which] will recur little more in this book. Its function was to provide a basis for a theory of kinds, and we can now begin to put it to that use. Figures 1b and 1c show, respectively, a group of geese and a group of swans, each represented in its own feature space, one which separates group members and thus permits their reidentification. These various creatures should, like the ducks in 1a, be thought of as the everyday kinds recognized by members of some social group or community, rather than as the scientific kinds developed by zoologists or ornithologists. Though the kinds of everyday life open the way to and operate in the same way as those of the natural historical sciences, the conditions required for their adequate function are vastly less stringent. As in the case of ducks, little is known about the differential features which permit these reidentifications, but facial markings are again likely to be prominent, and the initial difficulties in moving between these groups are probably like those humans experience in learning to reidentify members of different ethnic groups.

5. The presence of white space is required also by the weaker principle, developed in chapter 4, that the lifelines of distinct objects may not intersect.

Figure 1a

Figure 1b

Figure 1c

II

The division of biological creatures into kinds thus permits the reidentifications that endows them with objecthood, but the achievement of reidentification, even in principle, must then depend on the ability to identify the
kind to which an unknown creature belongs. Only when its kind is known
can the feature space needed for its reidentification be put in play. That

identification of kinds is, in turn, made possible by placing them in a hierarchy which can be entered at the top or some intermediate point and traced downward until a kind whose members are individual creatures is found. For the creatures currently under discussion, figure 2 will suggest what is involved.[6]

Note first that the hierarchy illustrated in figure 2 is fixed at top and bottom by the evolutionary and developmental patterns described in the preceding chapter. At the top are track makers; at the bottom the individual objects, or Aristotelian substances, which make tracks; both are covered by concepts that human infants appear to control by the end of their first year of life. Immediately below the topmost node, track makers are split into living and nonliving things, a fundamental division (not shown in the diagram) that has had to be learned, revised, and relearned repeatedly during both phylogenetic and ontogenetic development. Because nonliving track makers are artefacts, classified by their function rather than by their qualities, I shall largely ignore them until the next chapter. For the moment, attention is restricted to living track makers, the products of biological evolution. The hierarchies used to classify them probably provide the fundamental form on which all taxonomic hierarchies are modeled.

For a first look at the operation of the hierarchy, note that its various nodes are labeled with the names of features likely to be useful in discriminating between the kinds immediately below it. As was the case with face discrimination, nobody knows a great deal about what these features are, but it is plausible to suppose that, in the space where ducks, geese, and swans are discriminated, features like the ratio of head length to neck length and of body width to body length play a role. Color and adult size doubtless provide other dimensions, and there will be many more besides. If the individuals shown in figures 1a, 1b, and 1c are located in this space of differentiae, they can be expected to cluster in clumps like those shown in figure 3.

---

6. Contrary to everyday usage, the members of any higher-level kinds are, for simplicity, here taken to be the kinds lying immediately below it in the hierarchy rather than being individuals. My dog Fido is thus a member of the kind *dogs* but not of the kind *animals*. It is the class of all dogs that belongs to the kind *animals*. If the higher levels are allowed to have individuals as members, some more elaborate technique for distinguishing the bottom level is required. That level is, as will appear, the only one of which the members may change qualities with time.

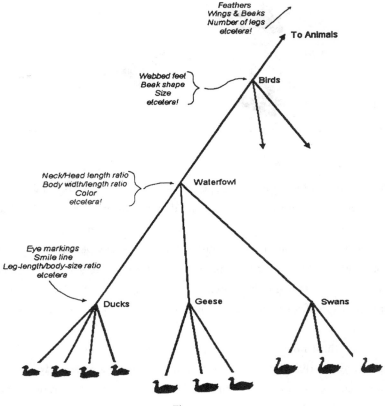

Figure 2

Now the empty space, which in figures 1a, 1b, and 1c expressed the prohibition of overlapping lifelines, appears between clusters and expresses the no-overlap principle for kinds. If it were not present, if there were, say, a creature that lay as close to geese as to swans, then the space would not be suitable for sorting these creatures into kinds. Other differentiae would be required, or else a resorting of the creatures which had previously unproblematically been ducks and geese.

Three characteristics of this space require special emphasis. First, as indicated by the prevalence of overlap between the creatures within the clumps in figure 3, the space that discriminates kinds is not generally suitable for separating individual lifelines. Occasionally, very unusual characteristics (birthmarks, damage to a leg or a wing) may permit discrimination

Figure 3

of a particular individual, but in general a lower-level space is required for the reidentification of individuals. Its dimensions are introduced at the node for the individual's kind as shown in figure 2. Second, though the space in which kind members cluster does not usually permit the tracing of individual lifelines, it must be rich enough to discriminate kinds at all stages of the lifelines of kind members. Otherwise, a particular individual might begin life as a member of one kind and later become a member of another.

The third noteworthy characteristic of the space in which kinds of creatures are differentiated is more complex: its full discussion would require consideration of standard difficulties about human motivations and actions, difficulties which can only be glanced at here. There are cases in which an object traversing a spatially and temporally continuous lifeline does become, slowly or quickly, a member of another kind. The most obvious examples are artefacts: cases in which the functions served by the object that emerges from a continuous change are quite different from those of the object which entered it. Think of the ice statue that becomes a tub of water, or the automobile that becomes a pile of scrap. The primary discussion of changes like these must wait for the discussion of artefacts in the next chapter. But living creatures also undergo functional changes of this sort at birth and again at death, in both cases changing the role they play in human practices. A duck egg is not a duck, though it may be food or a source for a new duck; a duck carcass is most likely to be food though it may, if a wild duck, be a candidate for taxidermy. At these points the creature changes kind, becoming or ceasing to be a living creature at all. A lifeline which extends through birth or, more obviously, death, would be one which permits a change of kind. Though it may be analyzable as the lifeline of a physical object, it cannot be the lifeline of any biological individual.[7]

How such lifelines are to be analyzed is problematic, for it raises controversial issues about human intentionality. But eggs, for example, are probably best treated initially as (ultimately separable) body parts, which after separation may become decaying physical objects, food, or living creatures. And dead ducks may be treated as carcasses (decaying physical objects), food, or trophies, the choice between [these] being heavily influenced by the mode of death. Though the choice between alternatives like these ultimately becomes straightforward, choice of the point in time at which the corresponding changes take place introduces an element of arbitrariness that sometimes has vast consequences for human behavior. Think of the passion invested in current debates about abortion and euthanasia: both involve the stipulation of the point during an observationally continuous change at which a change of kind takes place, a change that requires alteration in the acceptable responses to and responsibilities toward the altered object. Such choices are not avoidable, but they are also not determined by matters of fact. Later in this chapter I shall describe the nature of the resulting difficulties as intrinsic to the very concept of a kind.

We have been dealing, to this point, with the lowest-level kinds, those that contain individuals, and it is worth studying them further before turning to kinds at a higher level. Notice first that their members, though classified in a space with a relatively small number of dimensions, inherit properties from all the higher nodes one must pass through in reaching them from the track-makers node at the top. Ducks, for example, are alive because animals are, feathered because birds are, web-footed because waterfowl are, and so forth. This is the pattern of inheritance that permits the identification of the lowest-level feature space in which reidentification occurs. Given an unidentified creature, one enters the hierarchy of figure 2 at the lowest level to which one is confident the creature belongs (e.g., birds if it has wings

7. Note that we are back to Aristotle's distinction between *kinesis* and *metabole*, introduced in chapter 2. *Kinesis* is change along the interior of a lifeline, end points omitted: *metabole* extends the concept to include the lifeline's end points, the points where the object comes into being and passes away. Part of what the process of reidentification requires is a knowledge of the characteristics that mark the two end points.

and feathers) and then descends the hierarchy choosing, at each node, the branch for which it qualified by the differential features introduced at that node.

Seen as introducing a rich collection of differentiating features, the kinds to which these creatures belong are paradigms of what philosophers have, in recent years, been calling *natural kinds*.[8] They are "natural" because their members are, at any given time, among the observable elements of the natural environment. Human beings and other living creatures can [affect] the nature of the members of such kinds, but only by intervening in their developmental history: for example, by destroying them or their food supply or, in the case of human intervention, by breeding or genetic engineering.[9] Even if it becomes possible to create living creatures from inanimate materials, the forms of life that can be created will be limited by the kinds of material that nature makes available for their fabrication. Insofar as nature is "the given," it takes the form primarily of creatures and of the materials from which both living and nonliving objects are made. Geographical features—rivers, mountains, hills, lakes—provide a third set of natural kinds, and they will briefly enter the discussion of astronomy [in the] last section of this chapter. Taken together with creatures and materials, they may complete the roster of natural kinds.

To say that members of natural kinds are given is to say that their properties can be established by direct observation, independent of beliefs or theories about the causes of those properties and independent also of personal or social interest in their determination. Which properties are in fact observed and how closely the results of observation are subjected to critical scrutiny will, of course, be deeply influenced by interest and belief, and these are correspondingly important determinants of the rate and direction of cognitive development. But two people confronting the same creature or material can always—supposing they have normal sensory apparatus and speak the same

8. In this book the term *paradigm* is used only in its pre-Kuhnian form.

9. Use of the phrase *natural kinds* is not altogether uniform in the current philosophical literature, and it would therefore be possible to argue that human intervention, even if restricted to developmental history, makes its products artefactual rather than natural. But the next chapter will urge a more fundamental way of distinguishing natural from artefactual kinds, one that I am anticipating here: members of natural kinds can be classified by their observed properties; members of artefactual kinds cannot.

language—reach agreement about its observational properties. If one truthfully says it is green and another, also truthfully, that it is red, then either one
of them is color-blind or they use the language differently. Either or both of
these diagnoses could be correct: neural abnormalities and differences of
language can be subtle, difficult to discover. Color blindness, for example,
was unrecognized until early in the nineteenth century; significant examples
of the difficulties in locating differences in use of language are supplied by
chapter 2. But, confronted with disagreements about the observed properties of an individual object, members of a language/culture community have
urgent reasons to [resolve] them. If they could not reach agreement about
even the properties of observed objects, communication would break down,
thus undermining the foundation of the society that depended on it. The
point is crucial, and I shall be returning to it again and again.

The solidarity of a language community thus requires that its members be
able to achieve ultimate agreement about the observed properties of objects.
That generalization applies to objects in general, not just to living creatures
or members of natural kinds. But, since the observed properties of living
creatures change as they move along their lifelines, the reidentification which
makes them objects must be achieved in a feature space that clusters presentations of the same individual at times separated by some minimal time
interval Δt. Both the length of that interval and the dimensions of the space
within which it applies are functions of the kind to which the creature belongs.
It follows that to reidentify a living creature one must know the kind to which
that creature belongs. Later in this chapter we shall discover that no similar
condition applies to those natural kinds which, like materials, are not objects,
and in the next [chapter] we shall find that the same independence of kind characterizes the reidentification of artefactual objects (i.e., nonnatural kinds).

III

Members of natural kinds, whether creatures or materials, have one other
critically important characteristic, first emphasized by John Stuart Mill.[10]

---

10. J. S. Mill, *A System of Logic*, 8th ed. (New York: Harper and Brothers, 1881), 95–104,
406–10. In many other editions the material will be found in book I, chap. vii, §§ 3–6; III, xxii,
§§ 1–3.

Neither a finite set of observations nor their logical consequences can determine all the properties shared by members of a natural kind. Observationally, that is, members of natural kinds are inexhaustible. For Mill this characteristic provided the sometimes-elusive distinction between natural and artificial kinds. The members of an artificial kind, he said, were determined entirely by its definition. All yellow [plane] figures bounded by three straight lines belonged to the class yellow triangles, and conversely: qua members of that class there was nothing else to be observed about them. Natural kinds, Mill continued, also have definitions which single out their members in the same way. One or more subsets of the characteristics [that] members had been observed to share could supply necessary and sufficient conditions for membership in the kind; the particular subset that played that role could be chosen for its convenience. But no defining subset, however chosen, would exhaust the shared characteristics of kind members, even when enriched by its logical consequences.

Mill's discussion of the nature of natural kinds has played a decisive role in the development of my view on kinds in general. But it must already be apparent that his doctrine as stated will not do. Remember the discussion of the analytic/synthetic distinction at the beginning of chapter 3. A waterfowl could be black (or nonaquatic) and still a swan. Any of the apparently (and ordinarily) shared characteristics of the members of a natural kind could, in particular cases, be absent without changing the kind membership of a creature that lacked it. No feature shared by the previously known members of a kind is necessary for kind membership; no such feature can safely be made part of a definition of the kind.[11]

---

11. Mill, who knew the problems of biological classification, cannot have been unaware of the difficulties in finding a *definition* adequate to the task. But he saw no other way to distance his view of natural kinds from the Aristotelian notion, scorned since the seventeenth century, of kind membership being determined by a mysterious shared essence which made members of the kind what they were. That notion fits biological creatures particularly well, and Aristotle modeled all natural phenomena on the biological. For an especially useful account of Aristotle's concept of essence and of its biological sources, see Marjorie Grene, *A Portrait of Aristotle* (Chicago: University of Chicago Press, 1963), 78–85 and chap. 4. It cannot be accidental that, as the problem of meaning has become more and more acute during the last thirty years, an increasing number of the philosophers who deal with it have begun rapidly to rehabilitate the coupled notions of natural kind and essence. (The latter does not even appear in the index of the *Encyclopedia of Philosophy* published by Macmillan in 1967 [ed. Paul Edwards, 8 vols.].)

Features that kind members can be expected ordinarily to possess can, however, be dimensions in a space of differentiae intended to distinguish them from members of other kinds. Because the dimensions of that space are numerous, the absence of any particular feature will not prevent the anomalous creature from clustering with its fellows. If it does make difficulties—if the anomalous creature is located somewhere in what should be empty space—then further observations of it, and with the members of neighboring kinds, should provide an enriched space in which its kind membership is clear. If even that fails, then some of the kinds with which one has operated in the past may turn out not to be natural kinds at all, a possibility to which I shall return late in this chapter.

What I am suggesting, in short, is that the space in which members of a kind are distinguished from members of neighboring kinds is a space in which the location of the various cluster[s] brings with it expectations about the features of the members of the corresponding kinds. The individual who picks out kind members displays his or her expectations about the observable characteristics of members of that kind. If the lowest-level differential space is enriched by the features that descend from higher levels in the hierarchy, then the location of the clusters will also establish expectations about the characteristics that the various nearby kinds share: for example, feathers, wings, and webbed feet in the case of waterfowl. Some of these features may be more important, more salient, than others in establishing kind membership.[12] But none of them is individually necessary.

The essays collected in Stephen P. Schwartz, *Naming, Necessity, and Natural Kinds* (Ithaca, NY: Cornell University Press, 1977), suggest the suddenness with which that situation changed. In its index *essence, natural kinds,* and *natural kind terms* bulk large.

12. Mention of the special strength or salience of some features isolates a respect in which two-dimensional diagrams like figure 3 (or figures 1a, [1]b, and [1]c) are significantly misleading. They suggest neither the variety of the features which can constitute dimensions of the space, nor the vastly different breadth of the cluster in different dimensions. Some dimension[s] may be binary, feathered or not feathered; others may have a series of discrete values, [such as] the number of a creature's feet, from snail to centipede; still others—for example, size at maturity—vary continuously over a characteristic range. Features that are especially salient to the identification of kind members are provided by the dimensions along which the spread of a cluster is especially narrow. Salience is thus relative to the particular discriminations to be made: feathered/not-feathered is, for example, highly salient to discriminating birds from other animals, but not to discriminating between ducks, geese, and swans.

Generalization about their presence in members of a particular kind are always subject to exception; they are, that is, normic rather than nomic.

Where do such expectations come from? Or, to ask a closely related question, how does one know whether or not a given body of expectations selects members of a natural kind? To the first question, the answer is that one is taught to recognize kinds by someone who already knows; to the second, that one supposes the kind is natural because of the authority of one's teacher. What occurs during the interaction between learner and teacher is the transmission of the categories employed in the teacher's culture. Late in the 1960s, searching for a theory of meaning that would help explain the phenomena associated with incommensurability, I imagined a small child Johnny visiting a park with his father and learning during that visit to identify ducks, geese, and swans.[13] In outline the learning process went as follows. Father points to, say, a duck, and says, "Look, Johnny, there's a duck." A bit later Johnny points to another creature, saying, "Look, Daddy, another duck." Daddy looks but then says, "No Johnny, that's a goose." The process continues in this way, [with] swans entering the identification process along the way. Finally, after numerous successful and unsuccessful attempts, Johnny consistently reaches the same judgments as his father (teacher). He has learned to identify ducks, geese, and swans.

That dialogue is, of course, imagined, and it is hard to find direct evidence for (or against) the view it expresses, though I shall shortly supply just a bit. But let me first ask: Why [is] evidence so hard to find? I suspect the reason is that the relevant investigators—mostly developmental psychologists and philosophers—have failed to recognize the crucial difference between identification by characteristic features and [identification] by differentiae. Though most of them have abandoned the notion that classification and the meaning of kind terms require a generally accepted set of *characteristic* features, they have either abandoned features entirely in favor

---

13. "Second Thoughts on Paradigms," in *The Structure of Scientific Theories*, ed. Frederick Suppe (Urbana: University of Illinois Press, 1974), 459–99. That volume is the product of a conference held in 1969, to which the paper was presented in pretty much its published form. Readers will not be surprised to learn that the ducks, geese, and swans in the diagrams of this chapter (though not those diagrams themselves) first appeared there, drawn by my daughter, Sarah [Kuhn]. The paper is reprinted in my book of essays, *The Essential Tension*, 293–319.

of reference, or else they have continued to ask how features are employed and what the relevant features can be. In that process, the method[s] of characteristic features and of differentiae have appeared virtually equivalent. What has been overlooked is the fact that, though both depend on agreement between independent speakers or independent classifiers, what they require agreement about is very different.

There is, however, no reason to suppose that Johnny in my story is picking out ducks, geese, and swans in the same feature space as the one used by his teacher (father). In principle, the dimensions of the spaces they employ could be totally disjoint. What agreement requires is only that they apply the same labels to the same things, not that they do so in the same way. If two individuals (no longer child and father) agree about the referents of their shared terms, then each can learn from the other new ways of identifying those referents, both enriching their feature spaces in the learning process. What is required of their [feature] spaces individually is only that they cluster objects in the same way, produce empty space between the same collections of individuals: every duck must be closer to some other duck than any of them is to any swan or goose.

What must be shared is thus not differential feature spaces, but the structure exhibited in all of them by the kinds that can be found there. Kind terms do not attach to nature individually, one at a time. Johnny has not learned the concept of a duck (or the meaning of the word *duck*) until he has done the same for both geese and swans. What he must extract from the learning process is not a particular list of features, but rather some measure of similarity and difference that produces clusters of the members of each kind and puts empty space between them. In the process, he necessarily also acquires a similar measure of the arrangement of the various kinds: [a goose] and [a swan] are more like each other than either is like [a duck]. It is that arrangement that [we] *speak* of as "structure," and it is structure alone that must be shared by individuals who cluster the same individuals into the same kinds. That clustering is prerequisite to their ability to communicate unproblematically about the clustered creatures, and it is that communication which testifies to their sharing the relevant concepts and the same meanings. Ironically, prototype theory, the theory of meaning and concept formation most like my own, misses the need for this bit of local, non-Quinean holism. Like those who believe in characteristic features, prototype theorists conceive classification and meaning as attaching kinds to nature one at a time. The distance they are concerned with is that separating an individual from a prototypical example of the kind in which it should be placed.[14] No role is played by the distance to neighboring kinds.

Against that background, imagine the pleasure with which I greeted the following passage in a recent letter from a friend and sometimes collaborator resident in Zurich:

> Here is something our youngest, Alexander, did and does. He is now 22 months old. When we drove to the kindergarten which we do daily, he became quickly interested in the big moving things on the road, and we taught him—more or less unconsciously first—the terms *Lastwagen* [truck], *Bus*, and *Tram*. The training went exactly like your . . . old swan, goose, and duck example: no traits were mentioned, let alone definitions, only *Ja, das ist ein Lastwagen* [Yes, this is a truck] or *Nein, das ist kein Tram* [No, this isn't a tram.] His mastery of the three terms is now perfect, and he begins to get interested in other vehicles, too, especially motor bikes. But the parallel extends further. I don't know which traits he uses for identification, but they were surely visual in the beginning (because of head and eye movement and pointing). But a couple of days ago, he heard the sound of a *Bus* which is very typical indeed, in our garden where the traffic cannot be seen, and Alexander went *Bus! Bus!* So he extended the range of traits [used] to pick referents to acoustic ones, and that implies . . . that he has now, at least implicitly, empirical generalizations about buses at hand.[15]

Those observations can, of course, be explained in numerous other way[s]; the evidence they provide is not strong. But they do illustrate the way in which the initial viewpoint of the observer influences the observations that

14. For prototype theory, see Edward E. Smith and Douglas L. Medin, *Categories and Concepts* (Cambridge, MA: Harvard University Press, 1981).

15. Letter of April 27, 1994, from Paul Hoyningen-Huene in Zurich.

are made. With a change of viewpoint, it should be possible to uncover additional and stronger evidence.

Pending such evidence, let me develop a few more characteristics of the view of meaning and classification being developed here. The first has, in some respects, already been anticipated, especially in the last sentence of the preceding quotation. What are there called "empirical generalizations" are what I earlier called "expectations" or "normic generalizations." Learning to place individual creatures with members of their own kind is learning, simultaneously, much about the properties to be expected of members of that kind and of its neighbors. Further enriching the feature space initially acquired during training increases one's knowledge—one's store of normic generalizations—about the creatures located in it. What else could be expected from the process by which kinds and the meanings of their names are transmitted from one generation to the next? That process requires the presence both of someone who already knows, and real-world examples of the kinds of creatures he or she knows about. Pointing to those objects is crucial, both to isolate the example and to permit the recognition of mistakes.[16] Observation of the natural world is thus central not only to the initial development and evaluation of kind concepts but also to the process by which the results of inherited experience are retained in the culture which gave rise to them. That each generation can further enrich or otherwise change the kinds it inherits is the source of cultural evolution.

I use the phrase *retained in the culture*, because of another important aspect of Johnny's education in the park. What he acquired in interaction with his father was not only the ability to discriminate ducks, geese, and swans but also a command of features adequate to the discrimination. Some of the required features—perhaps those which enabled him to single out these waterfowl as birds—he had learned previously and brought to

16. It is often said that the operation of pointing or ostending is ambiguous in ways that defeat the purpose for which it [is] used here. One cannot tell, it is claimed, whether the inevitably imprecise arm-figure gesture is directed to, say, a swan, the swan's head, or a swan-shaped patch of white. But if context dictates, as it does in the present situation, that the gesture is directed to an object, then the white patch is excluded as are the parts that move together within the object's outline. Johnny, if he has learned to point at all, must intend to be pointing to a whole creature, a swan. That is among the lessons from the preceding chapter.

the park with him. They were already part of what I shall call his *vocabulary of features*, using *vocabulary* in the extended sense which does not require it to consist entirely of words.[17] That such a vocabulary must be acquired for reidentification of individuals was emphasized in the discussion of face recognition in the last chapter as well [as] in the discussion of reidentifying waterfowl in this one. The close relation between reidentifying and classifying makes it likely that a similar acquisition process should apply to the discrimination of kinds, and its doing so will prove to be the key to the historical examples presented in chapter 2.

That a significant vocabulary of features must be learned by interacting with the world in the company of someone who already knows a way of living there has numerous consequences. First, though the required features are learned by dealing with objects in the world, they are thereafter available for creating objects of myths and fictions. Horses and dogs and cats, like ducks and geese and swans, are real creatures because one can point to them and, with proper guidance, learn about the world in doing so. But what one learns about the world can be put to use in the imaginative creation of nonexistent creatures: winged horses, for example, centaurs, gorgons, and minotaurs. As one acquires also command of the kinds of emotion and personality recognized in one's culture—an acquisition that again requires both a guide and interaction with the world—one can deploy them to describe people who never lived and actions that never occurred. To claim that an object is or was real is to claim that it can or could [have been] both ostended and described. To claim that it is mythical or fictional is to claim that it can only be described; no person at any time or place could have pointed to it; what is real about it is only the features used to describe it, and they too had to be learned off the world.

This way of thinking about kinds and the identification of their members opens a possibility which is, I believe, regularly realized. Both kinds of objects and the features used to differentiate one of them from another are

17. Note that although there need not be a word for each of the features used in differentiating individuals or kinds, with the help of words and gestures, they can all be acquired through practice with a fellow speaker who makes use of them. Think of the friend who told me about the role of smile lines in reidentifying individual ducks or my talk of the ratio of neck length to body length in distinguishing ducks, geese, and swans.

cultural resources which each generation transmits, often in an enriched form, to the next. They shape the way members of a culture deal with their world as well as the tales they embody in myth and fiction. Those resources can, and I think clearly do, vary both from culture to culture, from discipline to discipline within a culture, and from one time to another within both cultures and disciplines.[18] Individual objects can be observed within any of these cultures, and its members may not, without risk to their culture, disagree about the presence of features their culture has equipped them to observe. But that compulsion does not apply to exchanges between member[s] of different cultures. Could they be said to disagree about the presence of a feature which only one of them had been equipped to observe? Rather than disagreement, misunderstanding will characterize their discourse. If they recognize misunderstanding, recourse is available to them, and I shall later have much to say about it. But the need for recourse is not easily recognized, and when recognized, it is not easily found, for it demands procedures not required for life within a culture. Those procedures are the ones illustrated by the examples in chapter 2, examples to which I shall be returning [in] the next chapter.

I am by now anticipating some of this book's central points, and they will need much further clarification and discussion, some of it in this chapter and some in later ones. But the way for them needs to be prepared, first with an additional anticipation and then, in the next two sections, by picking up a pair of topics introduced early in this chapter and repeatedly postponed since. As to the anticipation, the presumptive reason why people raised in different cultures differ in featural vocabularies is that their cultures here and there cluster objects into different kinds, and the features they acquire while learning to pick out kind members differ accordingly. In principle that need not be the case. Members of one culture can, I suggest, enrich their featural vocabulary with the features deployed by a different culture without jeopardy to their own, a sort of enrichment that provides a primary reason for the study of other cultures.[a] But the acquisition of

18. I am, of course, using terms like *culture* and *cultural* in a broad and perhaps idiosyncratic sense, and have found no way to avoid doing so. At least in complex societies, culture exists at various levels, and society members participate both in the larger culture and in a number of subcultures as well. Not all those subcultures are disciplines, but all disciplines are subcultures.

such an enlarged vocabulary without the instruction provided by interaction with (being taught by) another culture is highly unlikely. That is especially the case because the kind set of one culture, unlike the featural vocabulary which supports it, cannot be expanded to make room for the kinds employed by the other culture. Such an expansion would violate the no-overlap principle, destroying the integrity of one kind set or the other. That is the mismatch to which I have previously applied the term *incommensurability*.

These points are ones to which I shall be returning repeatedly, relating them in the process to the examples of chapter 2, and occasionally to other examples as well. This group of anticipations has now concluded. But before moving on, let me try to set aside a way of understanding incommensurability which has seriously inhibited understanding. Both featural vocabularies and kind sets are, I am claiming, cultural resources which vary from one culture or subculture to another. But between any pair of cultures, many kinds and many elements of the featural [vocabularies] must be shared. If they were [not], there would be no way to bridge the gap between them, no way for a member of one culture to learn the kind set and vocabulary of the other. It is not unimaginable that that should be the case, that a tribe should be discovered of which both language and behavior remained inscrutable despite sustained effort. But shared genetic endowment and overlap between environments makes it unlikely, and it is not clear that to speak of such a tribe as human would not involve a contradiction in terms. Incommensurability, as it is experienced in practice, is always a local phenomenon, restricted to one or more sets of interrelated concepts and the words that name them. Other concepts and terminology are shared and can be used to build the bridges that permit the learning process.

IV

To bring these anticipations more nearly within reach, look again at the hierarchy introduced in figure 2. As introduced, its function was to facilitate retrieval of differentiae relevant to reidentification by qualities of individual objects of a given kind. The possibility of such reidentification is what qualified these individuals as objects, the sorts of things that could be possessed, counted, exchanged, stolen, retrieved, and so on. And those qualifications are, in turn, what fitted them for fundamental roles in the varied practices of human societies. It is these practices, together with the requirement of shareable observations, that constrain the hierarchical structure of kinds, the structure through which society members interact both with each other with their shared world.

The constraints on a kind set are thus pragmatic, and in the remainder of this book I shall be increasingly suggesting that the only questions relevant when evaluating such a set concern its success in meeting its users' needs, including their need for shared observation. Needs vary, however, not only from culture to culture but also from one to another of the various subcultures found in all complex societies. Though there is much overlap, the constraints on the kind set of an agrarian society differ from those on the kind set of a society of hunters and gatherers. By the same token, the constraints on the kind set of a physicist differ from those on the kind set of a chemist and both from those of an engineer. Though all these cultures and subcultures have presumably evolved from a single stem, that evolution has been characterized by a continuing differentiation and specialization of their practices.

These remarks about hierarchy apply to objects of all kinds, artefactual as well as natural. But the features and hierarchies relevant to the former are very different from those for natural kinds, and I continue to restrict attention to the latter, for the moment to kinds of living things. For reasons to be considered below, a well-developed kind set for them would operate through differentiae at all levels below the node at which living and non-living objects separate. Every animal would then be closer to some other animal than to any nonanimal, and so on. No more than the lowest level of kinds—the level of ducks, geese, and swans—can the higher levels safely resort to necessary characteristic[s]. The black swans found in Australia illustrated the difficulties that can result for lowest-level kinds: the duck-billed platypus—an egg-laying creature that suckles its young—illustrates the same difficulties for a higher-level kind, mammals.

For purposes of everyday life, however, especially in a geographically restricted community, nothing much depends on achieving this ideal or even on all members' picking out the same set of creatures for the higher-level kinds. If ducks, geese, and swans are items in their practice (of eating, say, or of stuffing pillows and comforters), then they must be able to place these creatures in the same categories and, in some cases, to re-identify them (that's my duck, not yours). If irreconcilable disagreements at this level were at all usual, a constitutive practice of the community's would be in danger. But for higher-level categories the possibility of such disagreements rarely makes a difference. The primary [exception is] the categorical division between the living and [the] dead at the two ends of human lifelines, and that exception clearly illustrates the risks to community that result from disagreements affecting social practices. More typical is the category *animal* referred to above. Are fish animals? Are birds? And what about insects or worms? My *Random House Unabridged Dictionary* authorizes both a yes and a no answer to all these questions, and offers still another possibility: one standard English usage restricts animals to mammals. Each of these usages corresponds to a different way of drawing the hierarchical tree. In one, for example, birds, fish, animals, and insects descend from a single node. But all of these trees provide paths to the lower-level kinds needed for everyday life. Individuals tend to use more than one [hierarchical tree], depending on context. And whichever one they use, [it] leads them to the same identification in cases that make a difference. It would be hard to maintain that, for members of our geographically localized community, the differences between [hierarchical trees] [are disagreements] about matters of fact: that one is right, the others wrong. [Rather,] one [hierarchical tree] is perhaps more efficient when used for one purpose, another for another [purpose], but all lead to the same conclusions in everyday practice.

Life changes, however, and everyday life with it. The preceding discussion concerned the everyday life of a small, geographically isolated society, a society without specialized practices excepting perhaps a few due to gender. Now imagine that over time—enough time for many changes of generation—travelers explore lands at increasing distance, discover other human societies, and return bringing samples of plants and animals not previously encountered, some of them important to the practices of one or another of those societies. At this point problems arise for the original society, and they slowly accumulate with further exploration. Some of the trophies brought by returning travelers are simply exotic: the armadillo, perhaps, the turkey, or tobacco, and these clearly are members of a previously unknown kind. But others—especially but not exclusively from nearby regions—are quite like members of a locally familiar kind, and cluster with them using a standard set of differentiae. But they cluster even more closely with other specimens from the region where they were collected. Are these almost familiar individuals also members of a previously unknown kind or are they simply previously unknown varieties of a local kind, varieties which might have been produced at home by selective breeding or a change of environment? Are two kinds in question or only one?[19]

Failure to answer that question would threaten the viability of the society as previously constituted. Two sorts of threat are involved, the first short-term [and] the second long-term, and I examine them in order. Imagine a society divided between members who took the new specimens to be a variety of some familiar kind and members who took [them] to belong to a previously unknown kind, one requiring a new name and a new place in the taxonomy. Those who took them to be two varieties of a single kind would inevitably enrich their space of differentiae in ways that clustered the new specimens closely with previously familiar members of that kind, simultaneously increasing their distance from members of other kinds. Those who took them to be different kinds would enrich their space in ways that increased the separateness of their clusters. Though one cannot, except in highly artificial ways, cluster anything with anything, the alternatives just described are likely always to be available when there is ground for asking, Two kinds or one? If there had not been a strong resemblance from the start, the question would not have been asked.

The division I am hypothesizing can be described as a difference of belief about the new specimens, and in some sense, it is. But it is not an isolated belief, one about which the members of the society are free to differ.

19. This discussion of some problems of taxonomy owes much to discussions, thirty and more years ago, with my former colleague A. Hunter Dupree, and to his splendid book *Asa Gray, 1810–1888* (Cambridge, MA: Harvard University Press, 1959).

Belief in the integrity of its kind set had been somehow constitutive of that society, and the emerging difference will, if it continues, force the society's partial reconstitution. What will result over time is either the emergence of two separate societies or, more likely, the transition to a more complex society, one which legitimizes and institutionalizes the separation of the two groups. Since the difference between the parties has no merely factual basis, their members would continue to disagree about what belonged to a specified kind. As a result, a sale or exchange that satisfied a contract for members of one group might violate the agreement for members of the other, and there would be no factual grounds to provide a basis for negotiation. Or, members of one group could be held by members of the other to have violated dietary laws, and so on. But of all the practices infected by the division, the most central would be communication, a point to which I shall be returning again and again. Both parties would use the traditional name for the kind, and both would apply it to the traditional members. But one would apply it [to] the new specimens and the other would deny its applicability. The society which included both groups without differentiation would be in jeopardy. A statement one group thought true would be declared false by the other. And again, there would be no court of appeal. The problems I have been suggesting have, to this point, arisen not from matters of fact but more nearly from matters of usage dictated by something like taste or perhaps by the individual idiosyncrasies that had determined party membership in the first place.

Those are the short-term threats. They do not exist within either party taken alone, and they would disappear from the society as a whole if one usage or the other usage were enforced. Which one were chosen would make no *present* difference as long as usage were uniform. Enforcement is, however, a problem. Few societies have an authority licensed to compel assent in such matters of usage, and it is in any case unclear how effective enforcement could be achieved. There is, however, another and better way to manage the problem, one that emerges as a response to what I previously called the long-term threats posed by the new specimens. The traditional kind set used by members of the undifferentiated society was, I have suggested, a product of long-term experience with that society's world, a world which also played an essential role in the transmission of the society's kind set from one generation to the next. In that local world there was no place ready for the new specimens: they were unexpected, anomalous. If they were to be assimilated [to] the traditional kind set, [they] must, for the short-term reasons already discussed, be adjusted to fit. A further consideration increases the urgency of such adjustment: society members will be at risk until it is made. Confronted with the new specimens, they require answers to such questions as: Is the newly acquired plant poisonous? Is the newly discovered beast carnivorous? And so on. They need, in short, a basis for knowing what to expect of the new specimens and of others likely to be found in the future. They need, that is, some basis for reaching intelligent decisions about how to incorporate creatures not yet seen into future social practice. These long-term problems, unlike their short-term cousins, may make an important difference as the society develops over time. Here matters of fact are involved, but primarily future facts, not those presently at hand to members of either party. What is needed is something like a theory about what natural kinds can exist, something that will permit intelligent anticipations of the future. That need is one that an effective taxonomy can fulfill.

Such a taxonomy is asked is to provide not predictions but intelligent expectations, not nomic prescriptions but normic expectations. At least for natural kinds, however, expectations are what a taxonomy provides, and most of those expectations are embodied in and descend from higher-level kinds. For everyday purposes, many of the expectations that an improved hierarchy might yield are irrelevant. For purposes of everyday life, I've already indicated, it makes no difference whether birds are animals, or whether fish are. All three could, for example, descend from a single higher node. But, if they did, then the only expectations that were common to all three would descend from still-higher nodes like the one which separates living from nonliving creatures. Whereas, if fish and birds are both members of the higher-level kind *animals*, they will both share expectations that descend from that higher-level kind. Thus, if animals were taken to be mammals, fish and birds could not be subsets [of animals], for fish would then be expected to be warm-blooded and birds to be viviparous. Note that before the discovery of the platypus, no mammal was expected to lay eggs: the division between egg-laying creatures and those whose offspring were born alive had been placed above mammals in the hierarchy.

The structure of the hierarchy thus determines expectations not only about already familiar creatures but about those that may emerge in the future, and the need to make intelligent decisions about the latter places far-stronger constraints on hierarchical structure than those imposed by the needs of everyday life. If these expanded needs are to be met, then a hierarchical structure must embody a minimum number of arbitrary elements, and should instead respond to all available information about the kinds already known. Ideally, that is, upper-level kinds should, like those at the first level, be natural rather than artificial. Their members, too, should be determined by a space of differentiae in which all animals, say, are more like some other animals than they are like any member of a nonanimal kind. In such a space the problems posed by the discovery of black swans or of platypi might be easily resolved and could, in any case, be the subject of intelligent discussion. But finding such differentiae requires close study of (and sometimes also experiments upon) the considerable variety of creatures already known. Similarities and differences in the structure and function of internal organs are likely to be relevant: Which creatures have hearts and with how many chambers? Other relevant questions may concern the processes of reproduction and development: What creatures or plants can and cannot be paired to produce viable offspring? How are those offspring born and how nurtured? And those questions are only the beginning.

Because of their irrelevance to current concerns, the laity of the society can neither be expected nor asked to do the required studies. But, for the sake of the future, that same laity needs someone to do them. In the event, a group of specialists evolves to take responsibility for such tasks. All members of the larger society must know (or be able quickly to find out) who belongs to this group, for it is to them that society members confronted by an anomalous specimen will bring their questions. And it is society's recognized need for intelligent answers to such questions that gives the group [of specialists] its authority. Knowing that experts will be available when required, members of the larger society can get on with their lives in much the way they had before.

V

Further understanding, both of kinds and of hierarchy, can be obtained by brief consideration of another sort of natural kinds, one especially relevant to scientific development. These are the kinds of materials from which objects (no longer just natural kinds of objects) are made. Materials share three prominent characteristics with creatures: the role of differentiae in their identification; the role of hierarchy in locating the appropriate set of differentiae; and the role of observations about whose outcome members of the community must ordinarily agree. These I shall largely premise, concentrating instead on the differences and the consequences of the differences between these two sorts of natural kinds, differences which prove to be as noteworthy as the parallels. I shall consider four of them, all interrelated.

First, materials are not objects: they do not trace lifelines through space over time. Their names are mass nouns, not count nouns. Those names do not take the indefinite article ("*a* duck," but not "*a* gold"), and they have no plurals ("ducks" but not "golds"). Other noteworthy differences follow from these: "some gold" but not "some duck" provides a particularly prominent example.[20] The cleavage is fundamental and children master it at a relatively early stage of language development.[21]

A second distinction is closely related to this one. At the bottom of the hierarchy for living things are the individual creatures which are to be distributed among kinds. Some members of the community may be able to recognize more kinds than others (e.g., geese vs. kinds of geese), but for all of them the bottom level will be actual creatures, potential bearers of proper names, and they are not further subdivisible within the hierarchy. If members of this bottom level are divided at all it must be into parts: legs, wings, beaks, hearts, livers, and so on. Parts of creatures are, of course, objects in their own right, and they trace lifelines, but they are not living things and, once separated by force from their original owner—a separation without which they would not be objects—they are more appropriately described as artefactual than as natural kinds. In any case, other creatures have parts of the same kind—legs, livers, hearts, and so on—and

---

20. In fact, there are distinguishable circumstances under which the phrase *some duck* is grammatically acceptable. In the first, *some* is equivalent to *a* ("some duck has been into my vegetable patch"). In the second, *duck* is used as a mass noun, as in *duck meat* or *duck hunting*. Both are readily distinguished from the cases in which *some* indicates an unspecified part of the whole.

21. Nancy Soja, Susan Carey, and Elizabeth Spelke, "Ontological Categories Guide Young Children's Inductions of Word Meanings," *Cognition* 38, no. 2 (1991): 179–211.

these kinds too can be arranged hierarchically. But the hierarchy in which they find a place separates from the one for creatures at the node which divides living and nonliving things. The hierarchy for materials, on the other hand, bottoms out, not in individuals which belong to kinds but in kinds themselves—iron, water, wood, and so on. These too have parts, but unlike the parts of creatures, they are not directly observable until separated by human intervention (chemical or physical separation) from the material of which they are parts. In their separated state, they too are better described as artefactual than as natural, and the need for and difficulties of those separations again call forth one or more special communities of expert specialists.

That difference is the source of another. It will turn out to be no difference at all, but [the process of showing this to be the case] will disclose an essential characteristic of kinds, one not previously thematized. Creatures and other objects change over time. Kinds of material do not. People have believed that some everyday materials evolved from others, that lead, for example, was the most primitive of the metals and that it ripened in the earth, passing through such developmental stages as iron, copper, and gold. But that does not affect the constancy through time of the *kinds* of material: the evolving material was not gold until it could, by whatever means, be identified as gold: it was instead lead, or iron, or copper, or some other metal. That constancy does not, furthermore, imply a corresponding constancy in the means by which identification was achieved. Over historic time, as people have learned more about gold and developed more refined techniques for its identification, they have been able to show that objects once thought to be made of gold had, in fact, very little gold in their composition. But, though tests for and corresponding beliefs about gold may change with time, the kind named *gold* cannot itself change any more than the kind of figure named *triangle*. The place of gold within the kind set for metals must remain the same. Whatever set of differentiae are used in distinguishing metals, they must at all times preserve the empty space between them. If that were not the case, then an object once made of gold could continue to [be] made of gold while gradually changing into an object made of, say, iron. The kind named *gold* would then overlap in membership with the kind named *iron*. One would no longer be able to tell materials in the overlap apart. Like the similar difficulties in establishing the points at which life begins and ends, the problems of preserving the integrity of the kinds of metal demand a solution.

The no-overlap principle thus applies to kinds of material as well as to kinds of creature. Conversely, the arguments just given for the unchangeability of kinds of material applies equally to *kinds* of creatures. Indeed, if the identification of kind members is achieved by differentiae rather than [by] characteristic features, the no-overlap principle and the inalterability of kinds are equivalent. If overlap nevertheless occurs, that shows only that neither of the overlapping kinds was actually a kind at all, or at least not a natural kind. A similar principle for artefactual kinds will emerge in chapter 6, but the ground for it needs to be prepared.

A final difference between materials and creatures is the structure of the hierarchies in which they fall. The hierarchy of natural kinds of material is relatively simple: materials are, for example, divided between solids, liquids, and "airs" or gases; each of these may be further subdivided (liquids, for example, into oils, acids, and alkalis). But, in so far as these divisions are based entirely on properties observable without human intervention, the hierarchy has few levels, and there are only a few categories in each. But even for the purposes of everyday life, the hierarchy for creatures is far richer and more complex than the hierarchy for creatures. Though it does become extremely complex when experts are called upon, it continues to deal entirely with natural kinds of creatures. That is why botany, zoology, and other classificatory sciences are referred to as *natural* history. There are, of course, recognized experts on materials, too—chemists, physicists, and engineers—but they work downward from the basic level of materials, thus on the "parts" of materials; the kinds their work discloses are no longer natural but artefactual; and, like the parts of creatures, they belong in another hierarchy. Such artefactual hierarchies are postponed until the next chapter.

A last remark about the hierarchy for material concerns, not its difference from the hierarchy for creatures, but rather its independence of that hierarchy and of any hierarchy for objects. The hierarchy for creatures ascends to a terminus labeled *objects*; the corresponding node for materials is *materials*. There is no link between the two. For that reason, though the no-overlap principle applies within each hierarchy, it does not apply between them. The class of wooden things, for example, overlaps the class containing furniture, and so on, a complexity hinted at in the first pages of this chapter. There are other independent hierarchies, and similar overlaps can occur with each. The kind which contains dogs can overlap with both the class of males and the class of pets, and the latter two can overlap with each other. No-overlap is a condition only on individual hierarchies.

VI

I conclude this chapter by returning to a theme announced in the opening pages of chapter 1. Understanding the cognitive authority of science, I there suggested, would require resuscitation of the concept of incommensurability, a concept often seen as a threat to that authority. After much intervening preparation, that resuscitation can at last begin.

Both the reidentification of individuals and the identification of their kinds are, I have argued, most effectively achieved within a space of differential features. For reidentification, the space required must cluster previous presentations of a single individual at a distance from the clusters containing previous presentations of other individuals. For the identification of kind membership the required space must cluster members of a single kind at a distance from the members of other kinds. Candidates for identification or reidentification may then be assumed to belong in the cluster to which they lie closest. The technique is effective, of course, only so long as the world continues to behave as past experience has led community members to expect. If there is no cluster with which the candidate clearly belongs, special techniques are called for, but identification and reidentification in a space of differentiae is ordinarily unproblematic. That is the case, furthermore, even though members of the community may use spaces with quite different dimensions. (In principle the dimensions could be totally disjoint, though nothing so extreme is likely in practice.) They may not, of course, use any dimensions at all: satisfactory sets of differentiae must all yield the same clusters, but many sets will do that. To the extent that higher-level kinds are natural kinds, permissible sets of differentiae for them are constrained in the same way.

Now imagine two cultures (or two widely spaced stages in the development of a single culture) whose members used kind sets which, here and there, group the same bodies into different clusters. Both of them, for the sake of this first, somewhat artificial illustration, distinguish fish from animals.[b] But one of them groups whales and dolphins with cod and perch (by virtue of their aqueous home, the suitability of their shape to locomotion in it, and so on), the other groups them with otters and beavers (by virtue of their warm blood, method of reproduction, and so on). Obviously, the members of these two cultures have different concepts both of fish and of animals; the [feature spaces] they use to segregate the two are differently structured; and the terms used to refer to the two kinds have different meaning for members of each culture.

Under these circumstances it would create havoc to introduce into the conceptual vocabulary of one culture either the concepts or the names of the concepts used by the other to refer to fish and animals. The two cultures employ incommensurable kind sets; to enrich the vocabulary of one by introducing the terms *fish* or *animal* from the other would be to violate the no-overlap principle. If the traditional terms for animal in the two cultures are denoted by $fish_1$ and $fish_2$, both would name kinds that included whales: their inclusion within a single vocabulary would violate the no-overlap principle for kinds. The terms previously used by the members of each culture to denote fish and animals would, in the newly enriched language, lose their meanings.[22] And, in the absence of such enrichment, any statement about fish made in the language of one culture would be untranslatable into the language of the other.

That example is clearly contrived, and I shall therefore offer another. It is commonly said (in previous lives I have said it myself) that the Greeks believed the planets went around the Earth, while we believe the planets go around the Sun. That cannot be quite right, however, for the statement which compares the Greeks' belief with ours is incoherent. The Greeks'

---

22. I borrow the phrase from James Boyd White's *When Words Lose Their Meanings: Constitutions and Reconstitutions of Language, Character, and Community* (Chicago: University of Chicago Press, 1984), a book which from a very different source catches many of the problems that concern me. See also his *Justice as Translation: An Essay in Cultural and Legal Criticism* (Chicago: University of Chicago Press, 1990), especially its first three chapters.

conception of the planets (and the vocabulary they used to discuss them) was very different from ours. For us planets and stars are physical bodies which trace paths through space over time. For the Greeks they were more nearly features of physical bodies than bodies themselves. They did not, that is, have a motion of their own, but were carried by other bodies as our lakes, rivers, mountains, and other geographical features are carried by the Earth. Stars were features of the celestial sphere which rotated westward around the Earth each day. They, in turn, were divided into two species: the fixed stars (*aplanon astron*), which moved together, forever preserving their relative positions on the sphere; and the wandering stars (*planon astron*), which gradually fell behind the eastward motion of the fixed stars, thus moving westward through them. There were seven of these wandering stars: the Moon, Mercury, Venus, the Sun, Mars, Jupiter, and Saturn. Unlike the fixed stars, they did not twinkle, another characteristic that helped to distinguish them and which could be used without the delay, sometimes quite long, needed to determine whether a particular star had changed its relative position since a previous observation.

It is these wandering stars that ancient astronomers believed went around the Earth.[23] That cluster is different from ours, and it is formed in [a] space of differentiae with different dimensions. It thus includes the Sun, which for us is a star, and the Moon, which for us is not a planet but a satellite. And it excludes the Earth, which for us is a planet. Introducing the Greek concept or its name into our vocabulary would violate the no-overlap principle. We may choose between the two usages, but no working kind set can support them both simultaneously. That is what I had in mind when suggesting above that the standard way of comparing Greek and modern beliefs about the planets is incoherent. When we report that the Greeks believed the planets go around the Sun, we are attributing to them a concept of *planet* like our own, a concept which excludes the Sun by clustering it with the stars and which clusters the Moon with satellites, objects

23. Greek usage is somewhat equivocal with respect to the Sun and Moon. They are always spoken of as [planets] and grouped with the other wanderers when discussing their motion, their number, their distance from the Earth, and so on. But their obvious dissimilarities from other wanderers could not have been forgotten, and the texts frequently contain phrases like "the Sun, Moon, and planets." I am indebted to Noel Swerdlow for discussion of this point.

for which there was no room in the Greek cosmos. Greek statements about the planets and stars cannot be translated into our language.

There is, however, another way to capture what they had in mind. We can behave like anthropologists, acquiring their conceptual vocabulary, becoming vicarious (and very partial) members of their culture, and then using our own language, not to translate, but to teach their language to others. That is what I have been doing in the two examples of incommensurability just given. Unfortunately, however, in examples like these, language learning is so simple that its role is likely to be overlooked. In both cases, that is, the incommensurable parts of the two kind sets are populated by observable objects. One need only point to or name the objects in which the two kinds overlap, explaining how they are clustered by the other culture. That is not the case, however, if either the kind members or the features used to cluster them are not accessible to direct observation, as was the case with the examples of chapter 2. Because all of them involve artefactual rather than natural kinds, their discussion must be postponed to the next chapter. Meanwhile, let me use the case of natural kinds to suggest the difficulties [with] which incommensurability confronts standard views of scientific knowledge. If one cannot state two competing beliefs (or sets of belief[s]) in the same language, then one cannot compare them directly with observational evidence. That should not suggest that there are no good reasons why, over time, only one of them survives. Nor should it suggest that those reasons do not rest most fundamentally on observation. But it should suggest that the standard conception of a *choice* between the two on the basis of observational evidence cannot be quite right. Comparison requires simultaneous access to the things being compared, and that is here barred by the no-overlap principle. How that barrier is bridged is the subject of chapter 7. But we first need access to artefactual kinds.

September 24, 1995

CHAPTER 6

Practices, Theories, and Artefactual Kinds

The members of everyday artefactual kinds are, in the first instance, objects manufactured by living creatures, primarily human creatures. I think of tables and chairs, knives and forks, screwdrivers and can openers, houses and railway stations. Many of them are tools and most others can be thought of in that way. Like the naturally occurring objects discussed in the last chapter, they are directly observable and they make tracks through space over time. But in other respects the two are profoundly different, and a brief examination of their differences will isolate central characteristics that they share with such scientific concepts as force or weight, charge or insulator, gene or cell. Just as the taxonomic kinds which appear in the sciences arise from the natural kinds of everyday life and from the need for discourse about them, so the abstract kinds appearing in scientific theories plausibly arise from everyday artefactual kinds and from discourse about them.

I

With the notable exception of works of art and architecture, few artefacts can be reidentified by their observable qualities unless deliberately or accidentally labeled: by a serial number, for example, or by a scratch or missing chip. That they can be so labeled guarantees their status as objects, but in normal use such labels are rarely needed. On the contrary, it is an important characteristic of most artefacts that they be widely interchangeable;

for reasons shortly to be explained, many members of a given kind of arte-fact must ordinarily be available to replace each other without loss to the user. Those differences are paralleled by differences in the way the two sorts of objects are divided into kinds. In both cases, members of a kind must be identifiable by their observable properties; if they were not, it would be impossible to pick out an artefact of the kind one wanted. But those properties are not what groups them into kinds; rather, they are grouped together by their functions. (All can openers open cans, but they do not all display the same properties to someone unaware of the function they are to serve.) Unlike the members of natural kinds, observed properties pro-duce only the loosest sort of clustering (sometimes several loose clusters) and no role at all is played by differentiae, by the clustering of members of other kinds in the same space. The nature of artefacts is thus dual: as physi-cal objects they display observable properties, but it is their function, their place within a practice, that groups them into kinds.[1] Any member of the kind can serve that function, and the practice is facilitated by easy access to one or another member.

An example may clarify the distinction. It is sometimes difficult to tell a cup from a bowl. Cups usually have handles, bowls usually do not; most bowls are larger than most cups. But these two characteristics, though use-ful in our culture, are not always sufficient, and there are other cultures in which they are no use at all. Unlike the case of duck, geese, and swans, there is a *physical* continuum from cups to bowls: no empty space appears between them. But there is a difference in their functions: in our culture, cups are to drink from, bowls to eat from, and a well-equipped household contains both. It is immaterial that some vessels can serve either or both functions, though the efficient operation within an individual household may depend on its members' knowing to which use a given sort of vessel

---

1. For "duality" see William H. Sewell Jr., "A Theory of Structure: Duality, Agency, and Transformation," *American Journal of Sociology*, 98, no. 1 (1992): 1–29. Though the ideas it ad-vances have undergone a sea change on their way to me, this splendid article has provided badly needed guidance in the design of this chapter, especially of this section. That may be more plau-sible if I acknowledge that, until I was well into chapter 5, I thought of the distinction which here separates natural and artificial kinds as a division between what I called *taxonomic kinds* (exemplified by ducks, geese, and swans) and *singletons* (exemplified by mass, force, and weight).

is normally dedicated.[2] Eating is eating and drinking is drinking regardless of possible ambiguity in the physical form of the vessels employed.

Like the natural-kind members considered in the last chapter, most artefacts are observable physical objects, and disagreements within a culture about what those properties are places the culture itself at risk. But the functions which cluster artefacts into kinds are not observable in isolation. What must be observed instead is the whole practice within which a given artefact functions.[3] Such observations are part of a learning process which begins early in childhood (for our culture, toilet training and table manners provide examples), and the process continues into adulthood with increasing specialization by profession and discipline (e.g., law or medicine or chemical engineering). Participation of an adult practitioner is required throughout the learning process, and language is often (perhaps always) essential to the role the adult plays: "Cups are for drinking, bowls for eating" or "Don't eat off your knife, use your fork or spoon."[4] Furthermore, as those examples suggest, learning a practice requires learning several distinct functions together with the kinds of artefacts that serve them. In the last chapter I suggested that one cannot learn *duck* without simultaneously learning *goose* and *swan*. Now I suggest that one cannot learn *cup* without *bowl* or *fork* without *spoon*.[a] Artefacts and their functions are thus nodes in a practice, and the nodes are differentiated by relating their functions (usually through language) to those of other nodes which serve other functions. For artefactual kinds, unlike natural [kinds], there can be no talk of empty perceptual space or of nature's joints.

2. Asked by the cook to fetch the cups (or the bowls), a helper needs to know which vessels serve which function in the household. If they do not, several trips may be required to satisfy the request. There are activities, that is, with respect to which the household itself may be the relevant subculture. I owe this point (though not this way of expressing it) to Jehane Kuhn.

3. After functions have been learned and understood, it is often possible to identify what function an artefact serves from its physical characteristics. But such identifications are always risky and sometimes impossible. Think of a beanbag chair or a futon.

4. I am uncertain whether animals (nonlinguistic creatures) engage in practices, for practices have goals which practitioners must be able to retain while adjusting the practice to new circumstances. By that criterion, I am doubtful that the bee dance is a practice or that the division of labor within an ant colony suggests one. But the evidence is far from unequivocal, especially for the higher animals.

Though crudely and incompletely sketched, this view of physical artefacts and everyday practices will suffice for present purposes. Many of the features just examined will shortly reappear as features also of scientific theories, themselves central to the various specialized practices of the sciences. If more or less correct, the preceding remarks will suggest that much that is special about those practices emerged by refinement from everyday practices already in place. If rejected, those remarks should nevertheless clarify the discussion of the role of artefactual kinds in science. What follows does not depend on what precedes.

II

What primarily distinguishes natural kinds is, as the last chapter showed, that they are directly and immediately observable, thus the given. Which of their properties are in fact observed will vary from culture to culture, but members of any human culture can, if adequately motivated and trained, learn to recognize the properties which members of a different culture put to use in childhood. The properties which permit face recognition provided a particularly apposite example. As thought about natural kinds develops, unobservable entities are invoked to explain [their] observed properties or [the] behavior of their members. Think of *psyche*, the spirit or soul which animated living creatures and departed their bodies at death. Or think of the four Aristotelian elements—earth, air, fire, and water—which, unlike the observable materials which bore the same names, were all present in all material bodies but in different proportions. It is from roots like these that the scientific study of natural kinds develops, most notably [in] biology and chemistry.

There are, however, also sciences whose roots lie in the study not of natural, but of artefactual kinds. Physics is the most notable of these, and it originates in the study of matter in motion, both the concept of matter and that of motion being abstracted from the study of natural and artefactual kinds. Abstractions can, however, be made in different ways. The description of physics as the study of matter in motion fits Aristotle's physics just as well as it fits the physics of Galileo and Newton. But Aristotle's *motion* is not Galileo's or Newton's, and neither is Aristotle's *matter*. What

Aristotle abstracted as motion (*kinesis*) included change of qualities in general. What he abstracted as matter (at least in his *Physics*) was a substrate which endured through change of quality but which had always to be the bearer of some full set [of] qualities, whatever those qualities might be. For Galileo and Newton, on the other hand, motion was change of position alone, and matter was bare of all qualities except size, shape, position, and perhaps also weight, the so-called primary qualities. Neither of these ways of abstracting is properly described as right or wrong, true or false. What differs about them is their effectiveness as tools for practice in two quite different historical situations. That they are tools and that they came into existence through human agency is what makes it appropriate to group them with artefacts. Artefacts can, on this view, be intellectual as well as physical.

These two ways of abstracting correspond to different ways of disentangling what I called in chapter 4 the "Kantian components inextricably bound together in the neonate's tracking response." Anyone with sufficient vocabulary to describe an object in the process of altering its position as well as some of its properties can be taught either or both by verbal instruction. That is what I have just done. But the study of matter in motion involves other artefactual kinds—in the Newtonian case these include mass, force, and weight, all of which have nonequivalent Aristotelian versions. Unlike the members of natural kinds, the members of these artefactual kinds cannot be observed directly. Nor can they, unlike *matter* and *motion*, be abstracted one by one from observable members of either natural or artefactual kinds. They are, prototypically, what philosophers of science have labeled *theoretical terms* and contrasted with the terms in the *observation vocabulary* used to describe the members of both natural and artefactual kinds.[5]

---

5. The fit between the traditional conception of theoretical terms and the concept being developed here is quite close. But there is no equally full substitute in this book for the traditional notion of observation terms. Both natural-kind members and artefactual objects must be observable and members of any given culture must normally take for granted agreement about those observations, and there must be standard adjudication methods for the occasions when those expectations fail. But members of different cultures can have no similar cross-cultural expectations, although, with adequate training and experience, each can learn to make observations that are immediately accessible to the other.

How do individuals learn to use such terms? Through simultaneous interaction with someone who already knows how to use them and also with the world to which they apply. The learning process is thus a transmission process, from one generation to the next. It is also the process by which a new generation learns what kinds the world of the tribe which uses these terms contains.

BIBLIOGRAPHY

Agassi, Joseph. *Towards an Historiography of Science*. The Hague: Mouton, 1963.

Algra, Keimpe. "Concepts of Space in Classical and Hellenistic Greek Philosophy." PhD diss., Utrecht University, 1988. Published as *Concepts of Space in Greek Thought*. Philosophia Antiqua 65. Leiden: Brill, 1994.

Anscombe, G. E. M., and P. T. Geach. *Three Philosophers: Aristotle, Aquinas, Frege*. Ithaca, NY: Cornell University Press, 1961.

Aristotle. *Generation of Animals*. Translated by A. L. Peck. Loeb Classical Library 366. Cambridge, MA: Harvard University Press, 1942.

Aristotle. *On Sophistical Refutations. On Coming-to-Be and Passing Away. On the Cosmos*. Translated by E. S. Forster and D. J. Furley. Loeb Classical Library 400. Cambridge, MA: Harvard University Press, 1955.

Aristotle. *On the Heavens*. Translated by W. K. C. Guthrie. Loeb Classical Library 338. Cambridge, MA: Harvard University Press, 1939.

Aristotle. *Physics*. Edited by W. D. Ross. Translated by R. P. Hardie and R. K. Gaye. Vol. 2 of *The Works of Aristotle*. Oxford: Clarendon Press, 1930.

Aristotle. *Physics*. Translated by P. H. Wicksteed and F. M. Cornford. 2 vols. Loeb Classical Library 228, 255. Cambridge, MA: Harvard University Press, 1957.

Asquith, Peter D., and Henry E. Kyburg Jr., eds. *Current Research in Philosophy of Science*. East Lansing, MI: Philosophy of Science Association, 1979.

Asquith, Peter D., and Thomas Nickles, eds. *PSA 1982*. Vol. 2. East Lansing, MI: Philosophy of Science Association, 1983.

Bacon, Francis. *The Works of Francis Bacon*. Edited by James Spedding, Robert Leslie Ellis, and Douglas Denon Heath. Vol. 8, *Translations of the Philosophical Works*. New York: Hugh and Houghton, 1869.

Baillargeon, Renée. "Object Permanence in 3½- and 4½-Month-Old Infants." *Child Development* 23, no. 5 (1987): 655–64.

Baillargeon, Renée. "Representing the Existence and Location of Hidden Objects: Object Permanence in 6- and 8-Month-Old Infants." *Cognition* 23, no. 1 (1986): 21–41.

Baillargeon, Renée. "Young Infants' Reasoning about the Physical and Spatial Properties of a Hidden Object." *Cognitive Development* 2, no. 3 (1987): 179–200.

Benjamin, Walter. "The Task of the Translator." In *Illuminations*. Edited by Hannah Arendt. Translated by Harry Zohn, 69–82. New York: Harcourt Brace & World, 1968.

Bohr, Niels. *Causality and Complementarity*. Vol. 4 of *The Philosophical Writings of Niels Bohr*, edited by Jan Faye and Henry J. Folse. Woodbridge, CT: Ox Bow Press, 1998.

Bornstein, Marc H. "Perceptual Categories in Vision and Audition." In Harnad, *Categorical Perception*, 287–300.

Bower, T. G. R. *Development in Infancy*. 2nd ed. San Francisco: W. H. Freeman, 1982.

Brower, Reuben A., ed. *On Translation*. Cambridge, MA: Harvard University Press, 1959.

Brown, Theodore M. "The Electric Current in Early Nineteenth-Century French Physics." *Historical Studies in the Physical Sciences* 1 (1969): 61–103.

Bushneil, I. W. R., F. Sai, and J. T. Mullin. "Neonatal Recognition of the Mother's Face." *British Journal of Developmental Psychology* 7, no. 1 (1989): 3–15.

Butterfield, Herbert. *The Origins of Modern Science: 1300–1800*. London: G. Bell, 1949.

Butterfield, Herbert. *The Whig Interpretation of History*. London: G. Bell, 1931.

Carey, Susan. *The Origin of Concepts*. Oxford Series in Cognitive Development. Oxford: Oxford University Press, 2009.

Cheney, Dorothy L., and Robert M. Seyfarth. *How Monkeys See the World*. Chicago: Chicago University Press, 1990.

Clark, Eve V., ed. *The Proceedings of the Twenty-Sixth Annual Child Language Research Forum*. Stanford, CA: CSLI Publications, 1995.

Crombie, Alistair C. "Mechanistic Hypotheses and the Scientific Study of Vision: Some Optical Ideas as a Background to the Invention of the Microscope." In *Historical Aspects of Microscopy*, edited by S. Bradbury and G.L'E. Turner, 3–112. Cambridge: Heffer and Sons, 1967.

Davidson, Donald, and Gilbert Harman, eds. *Semantics of Natural Language*. 2nd ed. Dordrecht: Reidel, 1972.

de La Rive, Auguste Arthur. *Traité d'électricité théorique et appliquée*. Vol. 2. Paris: J.-B. Baillière, 1856.

Descartes, René. *Le Monde*. Edited by Charles Adam and Paul Tannery. Vol. 11 of *Œuvres de Descartes*. Paris: Léopold Cerf, 1909.

de Waard, Cornelis. *L'expérience barométrique: Ses antécédents et ses explications*. Thouars: Imprimerie Nouvelle, 1936.

Dewey, John. *Logic: The Theory of Inquiry*. New York: Henry Holt, 1938.

Dewey, John. "Propositions, Warranted Assertability, and Truth." *Journal of Philosophy* 38, no. 7 (1941): 169–86.

Dupree, A. Hunter. *Asa Gray, 1810–1888*. Cambridge, MA: Harvard University Press, 1959.

Edwards, Paul, ed. *Encyclopedia of Philosophy*. 8 vols. New York: Macmillan, 1967.

Eimas, Peter D., Joanne L. Miller, and Peter W. Jusczyk. "On Infant Speech Perception and the Acquisition of Language." In Harnad, *Categorical Perception*, 161–95.

Ereshefsky, Marc, ed. *The Units of Evolution: Essays on the Nature of Species*. Cambridge, MA: MIT Press, 1992.

Faraday, Michael. "Experimental Researches in Electricity." *Philosophical Transactions of the Royal Society of London* 122 (January 1832): 125–62.

Feyerabend, Paul K. "Explanation, Reduction, and Empiricism." *Minnesota Studies in Philosophy of Science* 3 (1962): 28–97.

Feyerabend, Paul K. "Putnam on Incommensurability." *British Journal for the Philosophy of Science* 38, no. 1 (1987): 75–92.

Field, Tiffany M., Debra Cohen, Robert Garcia, and Reena Greenberg. "Mother-Stranger Face Discrimination by the Newborn." *Infant Behavior and Development* 7, no. 1 (1984): 19–25.

Fodor, Jerry A., Merrill F. Garrett, Edward C. T. Walker, and Cornelia H. Parkes. "Against Definitions." *Cognition* 8, no. 3 (1980): 263–367.

Grene, Marjorie. *A Portrait of Aristotle*. Chicago: University of Chicago Press, 1963.

Griffin, Donald R. *Animal Minds*. Chicago: University of Chicago Press, 1992.

Gutting, Gary. "Continental Philosophy and the History of Science." In Olby et al., *Companion to the History of Modern Science*, 127–47.

Hacking, Ian. "Language, Truth, and Reason." In *Rationality and Relativism*, edited by Martin Hollis and Steven Lukes, 48–66. Cambridge, MA: MIT Press, 1982.

Hacking, Ian. *Representing and Intervening: Introductory Topics in the Philosophy of Natural Science*. Cambridge: Cambridge University Press, 1983.

Hacking, Ian. *Why Does Language Matter to Philosophy?* Cambridge: Cambridge University Press, 1975.

Hacking, Ian. "Working in a New World: The Taxonomic Solution." In *World Changes: Thomas Kuhn and the Nature of Science*, edited by Paul Horwich, 275–310. Cambridge, MA: MIT Press, 1993.

Hanson, Norwood Russell. *Patterns of Discovery*. Cambridge: Cambridge University Press, 1958.

Harnad, Steven, ed. *Categorical Perception: The Groundwork of Cognition*. Cambridge: Cambridge University Press, 1987.

Held, Richard. "Binocular Vision: Behavior and Neuronal Development." In Mehler and Fox, *Neonate Cognition*, 37–44.

Hempel, Carl G. *Philosophy of Natural Science*. Englewood Cliffs, NJ: Prentice Hall, 1966.

Hiley, David R., James F. Bohman, and Richard Shusterman, eds. *The Interpretive Turn*. Ithaca, NY: Cornell University Press, 1991.

Hirsch, Eli. *The Concept of Identity*. New York: Oxford University Press, 1982.

Hollis, Martin, and Steven Lukes, eds. *Rationality and Relativism*. Cambridge, MA: MIT Press, 1982.

Horwich, Paul, ed. *World Changes: Thomas Kuhn and the Nature of Science*. Cambridge, MA: MIT Press, 1993.

Hull, David. *Science as Process: An Evolutionary Account of the Social and Conceptual Development of Science*. Chicago: University of Chicago Press, 1988.

Jakobson, Roman. "On Linguistic Aspects of Translating." In Brower, *On Translation*, 232–39.

Jammer, Max. *The Conceptual Development of Quantum Mechanics*. New York: McGraw Hill, 1966.

Jevons, W. Stanley. *The Principles of Science: A Treatise on Logic and Scientific Method*. 1874. Reprint, New York: Dover, 1958.

Johnson, Mark H., and John Morton. *Biology and Cognitive Development: The Case of Face Recognition*. Oxford: Blackwell, 1991.

Kellman, Philip J., and Elizabeth S. Spelke. "Perception of Partly Occluded Objects in Infancy." *Cognitive Psychology* 15, no. 4 (1983): 483–524.

Kellman, Philip J., Elizabeth S. Spelke, and Kenneth R. Short. "Infant Perception of Object Unity from Translatory Motion in Depth and Vertical Translation." *Child Development* 57, no. 1 (1986): 72–86.

Kox, A. J., ed. *The Scientific Correspondence of H. A. Lorentz*. Vol. 1. New York: Springer, 2009.

Koyré, Alexandre. *Études galiléennes*. 3 vols. Paris: Hermann, 1939.

Kripke, Saul. "Naming and Necessity." In *Semantics of Natural Language*, 2nd ed., edited by Donald Davidson and Gilbert Harman, 253–355. Dordrecht: Reidel, 1972. Reprinted as *Naming and Necessity*, Harvard University Press, 1980.

Krüger, Lorenz, Lorraine J. Daston, and Michael Heidelberger, eds. *The Probabilistic Revolution*. Vol. 1, *Ideas in History*. Cambridge, MA: MIT Press, 1987.

Kuhn, Thomas S. "Afterwords." In *World Changes: Thomas Kuhn and the Nature of Science*, edited by Paul Horwich, 314–19. Cambridge, MA: MIT Press, 1993. Reprinted in Kuhn, *Road Since Structure*, 224–52.

Kuhn, Thomas S. *Black-Body Theory and the Quantum Discontinuity, 1894–1912*. 2nd ed. Chicago: University of Chicago Press, 1987. First published 1978 by Oxford University Press.

Kuhn, Thomas S. "Commensurability, Comparability, Communicability." In *PSA 1982*, vol. 2, edited by Peter Asquith and Thomas Nickles, 669–88. East Lansing, MI: Philosophy of Science Association. Reprinted in Kuhn, *Road Since Structure*, 33–57.

Kuhn, Thomas S. *The Essential Tension: Selected Studies in Scientific Tradition and Change*. Chicago: University of Chicago Press, 1977.

Kuhn, Thomas S. "The Natural and the Human Sciences." In *The Interpretive Turn: Philosophy, Science, Culture*, edited by David R. Hiley, James F. Bohman, and Richard Shusterman, 17–24. Ithaca, NY: Cornell University Press, 1991. Reprinted in Kuhn, *Road Since Structure*, 216–23.

Kuhn, Thomas S. "Possible Worlds in History of Science." In Allén, *Possible Worlds in Humanities, Arts, and Sciences*, 9–32.

Kuhn, Thomas S. "Rationality and Theory Choice." *Journal of Philosophy* 80, no. 10 (1983): 563–70. Reprinted in Kuhn, *Road Since Structure*, 208–15.

Kuhn, Thomas S. "Revisiting Planck." *Historical Studies in the Physical Sciences* 14, no. 2 (1984): 231–52. Reprinted as the afterword to the paperback edition of Kuhn, *Black-Body Theory*, 349–70.

Kuhn, Thomas S. *The Road Since Structure: Philosophical Essays, 1970–1993, with an Autobiographical Interview.* Edited by James Conant and John Haugeland. Chicago: University of Chicago Press, 2000.

Kuhn, Thomas S. "Second Thoughts on Paradigms." In *The Structure of Scientific Theories*, ed. Frederick Suppe, 459–99. Urbana: University of Illinois Press, 1974. Reprinted in Kuhn, *Essential Tension*, 293–319.

Kuhn, Thomas. *The Structure of Scientific Revolutions.* 2nd ed. Chicago: University of Chicago Press, 1970. First published 1962 by the University of Chicago Press.

Kuhn, Thomas S. "What Are Scientific Revolutions?" In *The Probabilistic Revolution*. Vol. 1, *Ideas in History*, edited by Lorenz Krüger, Lorraine Daston, and Michael Heidelberger, 7–22. Cambridge, MA: MIT Press, 1987. Reprinted in Kuhn, *Road Since Structure*, 13–32.

Landau, Terry. *About Faces.* New York: Anchor Books, 1989.

Laudan, Larry. "Historical Methodologies: An Overview and Manifesto." In *Current Research in Philosophy of Science*, edited by Peter D. Asquith and Henry E. Kyburg Jr., 40–54. East Lansing, MI: Philosophy of Science Association, 1979.

Liddell, Henry George, and Robert Scott. *A Greek-English Lexicon.* 9th ed. 2 vols. Revised and augmented throughout by Sir Henry Stuart Jones, with the assistance of Roderick McKenzie. Oxford: Clarendon Press, 1940.

Locke, John. *An Essay Concerning Human Understanding.* 1689. Edited by Peter H. Nidditch. Oxford: Oxford University Press, 1979.

Lyons, John. *Semantics.* 2 vols. Rev. ed. Cambridge: Cambridge University Press, 1984.

Markman, Ellen M. *Categorization and Naming in Children: Problems of Induction.* Cambridge, MA: MIT Press, 1989.

Marr, David. *Vision: A Computational Investigation into the Human Representation and Processing of Visual Information.* San Francisco: W. H. Freeman, 1982.

Maurer, Daphne, and Philip Salapatek. "Developmental Changes in the Scanning of Faces by Young Infants." *Child Development* 47, no. 2 (1976): 523–27.

Mayr, Ernst. "Biological Classification: Toward a Synthesis of Opposing Methodologies." *Science* 214, no. 4520 (1981): 510–16.

McMullin, Ernan, ed. *Construction and Constraint: The Shaping of Scientific Rationality.* Notre Dame, IN: University of Notre Dame Press, 1988.

Mehler, Jacques, and Robin Fox, eds. *Neonate Cognition: Beyond the Blooming Buzzing Confusion.* Hillsdale, NJ: Erlbaum, 1985.

Meltzoff, Andrew N., and M. Keith Moore. "Early Imitation within a Functional Framework: The Importance of Person Identity, Movement, and Development." *Infant Behavior and Development* 15, no. 4 (1992): 479–505.

Mill, John Stuart. *A System of Logic, Ratiocinative and Inductive, Being a Connected View of the Principles of Evidence and the Methods of Scientific Investigation.*

Vols. 7–8 of *The Collected Works of John Stuart Mill*, edited by John M. Robson. London: Routledge and Kegan Paul, 1963–74.

Millikan, Ruth Garrett. *Language, Thought, and Other Biological Categories: New Foundations for Realism*. Cambridge, MA: MIT Press, 1984.

Nida, Eugene. "Principles of Translation as Exemplified by Bible Translating." In Brower, *On Translation*, 11–31.

Olby, R. C., G. N. Cantor, J. R. R. Christie, and M. J. S. Hodge, eds. *Companion to the History of Modern Science*. London: Routledge, 1990.

Peirce, Charles S. "How to Make Our Ideas Clear." *Popular Science Monthly*, January 1878.

Piaget, Jean, and Bärbel Inhelder. *The Child's Conception of Space*. Translated by F. J. Langdon and J. L. Lunzer. New York: Norton, 1967.

Pickering, Andrew, ed. *Science as Practice and Culture*. Chicago: University of Chicago Press, 1992.

Putnam, Hilary. *Reason, Truth, and History*. Cambridge: Cambridge University Press, 1981.

Quine, W. V. O. *From a Logical Point of View*. Cambridge, MA: Harvard University Press, 1953.

Quine, W. V. O. *Word and Object*. Cambridge, MA: MIT Press, 1960.

Reichenbach, Hans. *Experience and Prediction: An Analysis of the Foundations and the Structure of Knowledge*. Chicago: University of Chicago Press, 1938.

Repp, Bruno H., and Alvin M. Liberman. "Phonetic Category Boundaries are Flexible." In Harnad, *Categorical Perception*, 89–112.

Rosen, Stuart, and Peter Howell. "Auditory, Articulatory, and Learning Explanations in Speech." In Harnad, *Categorical Perception*, 113–160.

Russell, Bertrand. *The Problems of Philosophy*. London: Williams and Norgate, 1912.

Schagrin, Morton L. "Resistance to Ohm's Law." *American Journal of Physics* 31, no. 536 (1963): 536–47.

Scheffler, Israel. *Science and Subjectivity*, 2nd. ed. Indianapolis: Hackett, 1982.

Schwartz, Stephen P., ed. *Naming, Necessity, and Natural Kinds*. Ithaca, NY: Cornell University Press, 1977.

Sewell, William H., Jr. "A Theory of Structure: Duality, Agency, and Transformation." *American Journal of Sociology* 98, no. 1 (1992): 1–29.

Slaughter, Mary M. *Universal Languages and Scientific Taxonomy in the Seventeenth Century*. Cambridge: Cambridge University Press, 1982.

Smith, Edward E., and Douglas L. Medin. *Categories and Concepts*. Cambridge, MA: Harvard University Press, 1981.

Soja, Nancy N., Susan Carey, and Elizabeth S. Spelke. "Ontological Categories Guide Young Children's Inductions of Word Meanings." *Cognition* 38, no. 2 (1991): 179–211.

Sorabji, Richard. *Matter, Space, and Motion: Theories in Antiquity and Their Sequel*. Ithaca, NY: Cornell University Press, 1988.

Spelke, Elizabeth S. "Perception of Unity, Persistence, and Identity: Thoughts on Infants' Conceptions of Objects." In Mehler and Fox, *Neonate Cognition*, 89–113.

Spelke, Elizabeth S. "Principles of Object Perception." *Cognitive Science* 14, no. 1 (1990): 29–56.

Spelke, Elizabeth S., Karen Breinlinger, Janet Macomber, and Kristen Jacobson. "Origins of Knowledge." *Psychological Review* 99, no. 4 (1992): 605–32.

Steiner, George. *After Babel: Aspects of Language and Translation*. London: Oxford University Press, 1975.

Sture, Allén, ed. *Possible Worlds in Humanities, Arts, and Sciences: Proceedings of Nobel Symposium 65*. Berlin: Walter de Gruyter, 1989.

Suppe, Frederick, ed. *The Structure of Scientific Theories*. Urbana: University of Illinois Press, 1974.

Sutton, Geoffrey. "The Politics of Science in Early Napoleonic France: The Case of the Voltaic Pile." *Historical Studies in the Physical Sciences* 11, no. 2 (1981): 329–66.

Taylor, Charles. "Interpretation and the Sciences of Man." *Review of Metaphysics* 25, no. 1 (1971): 3–51. Reprinted in his *Philosophy and the Human Sciences*. Vol. 2 of *Philosophical Papers*, 15–57. Cambridge: Cambridge University Press, 1985.

Ullman, Shimon. *The Interpretation of Visual Motion*. Cambridge, MA: MIT Press, 1979.

Volta, Alessandro. "On the Electricity Excited by the Mere Contact of Conducting Substances of Different Kinds." *Philosophical Transactions of the Royal Society* 90 (1800): 403–31.

Werner, Heinz. *Comparative Psychology of Mental Development*. Rev. ed. Chicago: Follett, 1948.

White, James Boyd. *Justice as Translation: An Essay in Cultural and Legal Criticism*. Chicago: University of Chicago Press, 1990.

White, James Boyd. *When Words Lose their Meanings: Constitutions and Reconstitutions of Language, Character, and Community*. Chicago: University of Chicago Press, 1984.

Wiggins, David. *Sameness and Substance*. Cambridge, MA: Harvard University Press, 1980.

Wittgenstein, Ludwig. *Philosophical Investigations*. 4th ed. Edited by P. M. S. Hacker and Joachim Schulte. Translated by G. E. M. Anscombe, P. M. S. Hacker, and Joachim Schulte. Oxford: Wiley-Blackwell, 2009. First published, in G. E. M. Anscombe's translation, 1953 by Blackwell (Oxford).

Wittgenstein, Ludwig. *Remarks on Colour*. Edited by G. E. M. Anscombe. Translated by Linda L. McAlister and Margarete Schättle. Berkeley: University of California Press, 1977.

Xu, Fei, and Susan Carey. "Infants' Metaphysics: The Case of Numerical Identity." *Cognitive Psychology* 30, no. 2 (1996): 111–53.

Xu, Fei, Susan Carey, and Nina Quint. "The Emergence of Kind-Based Object Individuation in Infancy." *Cognitive Psychology* 49, no. 2 (2004): 155–90.

Xu, Fei, Susan Carey, Kyra Raphaelidis, and Anastasia Ginzbursky. "12-month-old Infants Have the Conceptual Resources to Support the Acquisition of Count Nouns." In *Proceedings of the Twenty-Sixth Annual Child Language*

Research Forum, edited by Eve V. Clark, 231–38. Stanford, CA: CSLI Publications, 1995.

Xu, Fei, Susan Carey, and Jenny Welch. "Infants' Ability to Use Object Kind Information for Object Individuation." *Cognition* 70, no. 2 (March 1, 1999): 137–66.

NOTES

Editor's Introduction

1. *The Structure of Scientific Revolutions* (Chicago: University of Chicago Press, 1962), henceforth *Structure*. The second revised edition, to which all subsequent references will be made, was published in 1970, also by Chicago.

2. This is clearly the case with the general philosophy of science, but Kuhn's work was also influential in the philosophy of language, epistemology, history and sociology of science, science studies, and other, more distant fields. For example, the terms *paradigm* and *incommensurability* are ubiquitous in contemporary academia; their multiple usages are all inspired by *Structure*.

3. Most of these papers are collected and posthumously published in *The Road Since Structure: Philosophical Essays, 1970–1993, with an Autobiographical Interview*, ed. James Conant and John Haugeland (Chicago: University of Chicago Press, 2000).

4. Giving the reader the necessary information and context without imposing my own interpretation and evaluation of Kuhn's work was a delicate balancing act. Although explication and interpretation in philosophy inevitably go together and infuse each other, there are ways to privilege one over the other. When in doubt, I always opted for editorial and interpretive restraint. For example, I decided to offer only a few clarificatory endnotes to the primary texts, and so I never treated the endnotes as a space to philosophically engage with Kuhn's views. (For more information about editorial decisions, see my editor's note.)

5. See especially "Possible Worlds in History of Science," "The Road since *Structure*," and "The Trouble with the Historical Philosophy of Science," reprinted as chapters 3–5 in *Road Since Structure*.

6. Thomas S. Kuhn Papers, MC 240, box 23, Institute Archives and Special Collections, Massachusetts Institute of Technology, Cambridge, Massachusetts

(henceforth IASC MIT). However, the Shearman Lectures, published here, offer a more developed version of the same set of ideas.

7. These are kept in the Thomas S. Kuhn Papers, MC 240, box 23, IASC MIT.

8. Thomas S. Kuhn Papers, MC 240, box 22, IASC MIT.

9. I received from the University of Chicago Press, with the permission of Kuhn's literary executors, Kuhn's notes for *The Plurality of Worlds*, together with the last version of the book manuscript.

10. Some of Kuhn's notes highlight the problems that he needs to resolve before considering the chapter complete; others suggest ideas that need to be developed and incorporated. Some notes, however, seem to be tangential to the chapter with which they are filed.

11. Sadly, Jehane Kuhn died in 2021. Kuhn's daughter, Sarah Kuhn, continues as his literary executor. The title of the transcripts reads "Interviews with Tom Kuhn, June 1996." The interviews were conducted by James Conant and John Haugeland, transcribed by Joan Wellman, and proofread by John Haugeland.

12. "They would not go into the archive under any circumstances," he said. "Interviews with Tom Kuhn," 142.

13. Brandeis University, May 30, 1984; University of Minnesota, October 21, 1985; Tokyo, May 2, 1986.

14. I was not able to discover why Kuhn thought the lecture would be published in *Synthèse*, nor why it was not published there.

15. Thomas S. Kuhn, "Rekishi Shosan toshite no Kagaku Chishiki" [Scientific knowledge as historical product], trans. Chikara Sasaki and Toshio Hakata, *Shisō* [Thought] 8, no. 746 (August 1986): 1–18. I thank Shinko Kagaya for her invaluable help in obtaining a copy of the Japanese publication. The text published here is from the original English version, kept in the Thomas S. Kuhn Papers, MC 240, boxes 23–24, IASC MIT.

16. The central ideas of the Notre Dame Lectures were reworked in "What Are Scientific Revolutions?" and "Commensurability, Comparability, and Communicability" (both reprinted in *Road Since Structure*, as chaps. 1 and 2, respectively), as well as in the Thalheimer and Shearman Lectures.

17. "Scientific Development and Lexical Change," Thalheimer Lectures, Johns Hopkins University, November 12–19, 1984. (Thomas S. Kuhn Papers, MC 240, box 23, IASC MIT.) The Thalheimer Lectures have been published in Spanish translation: *Desarrollo científico y cambio de léxico: Conferencias Thalheimer*, ed. Pablo Melogno and Hernán Miguel, trans. Leandro Giri (Montevideo, Uruguay: ANII/UdelaR/SADAF 2017). The first of the four lectures is discussed in print by Pablo Melogno; see his "The Discovery-Justification Distinction and the New Historiography of Science: On Thomas Kuhn's Thalheimer Lectures," *HOPOS: The Journal of the International Society for the History of Philosophy of Science* 9, no. 1 (Spring 2019): 152–78.

18. Kuhn's Shearman Lectures aroused such interest that there was no lecture room at University College London big enough to accommodate everyone. Kuhn distributed copies of the Shearman Lectures to some members of the

audience waiting in the corridors. These copies were afterward circulated without Kuhn's permission. "Interviews with Tom Kuhn," 111.

19. Ian Hacking, "Working in a New World: The Taxonomic Solution," in *World Changes: Thomas Kuhn and the Nature of Science*, ed. Paul Horwich, 275–310 (Cambridge, MA: MIT Press, 1993); Jed Z. Buchwald and George E. Smith, "Thomas S. Kuhn, 1922–1996," *Philosophy of Science* 64 no. 2 (1997): 361–76.

20. "Afterwords," in Horwich, *World Changes*, 311–41; reprinted as chap. 11 in *Road Since Structure*.

21. In writing each of the lecture series (Notre Dame, Thalheimer, and Shearman), Kuhn hoped to create a draft of the book, but he felt that he never succeeded. He thought that the Shearman Lectures came "very close," but he still did not want them published ("Interviews with Tom Kuhn," 50). It seems that he was particularly dissatisfied with a version of Kantian idealism that he endorsed in the Shearman Lectures (61, 81).

22. This is also how James A. Marcum refers to the unfinished book in his *Thomas Kuhn's Revolution* (London: Continuum, 2005), 25, 126.

23. David Lewis, *On the Plurality of Worlds* (Oxford: Blackwell, 1986).

24. The concern with similarity to Lewis's title appears on p. 92 of "Interviews with Tom Kuhn."

25. Although some of the book's central ideas date as far back as "Second Thoughts on Paradigms"—first published in *The Structure of Scientific Theories*, ed. Frederick Suppe, 459–82 (Urbana: University of Illinois Press, 1974); reprinted as chap. 12 in *The Essential Tension: Selected Studies in Scientific Tradition and Change* (Chicago: The University of Chicago Press, 1977)—Kuhn's work on the actual manuscript started only after he completed the Shearman Lectures.

26. Presentist case studies are generally simplified and completely decontextualized. They are used to make a particular point, usually a methodological one, that matters to present-day audiences. Presentism treats history of science as a series of stepping-stone achievements that straightforwardly lead to contemporary science. Seen through the distorting lens of our current beliefs, past scientists appear rational and insightful only if they can be described as precursors to contemporary work; otherwise their beliefs and practices are set aside as misguided, irrational, or even obstinately opposed to progress. Kuhn rightly thought that this is not how responsible historical research should be done. Controversially, at the time when *Structure* was published, he also believed that the philosophy of science cannot get off the ground without careful historical recovery of past science.

27. Kuhn was satisfied with the formulation he offered there. He said that the main point of the epilogue is "virtually liftable out of the last part of the Shearman Lectures" ("Interviews with Tom Kuhn," 65).

28. He sometimes whimsically referred to his last book as "the grandson of *Structure*," saying that "the son of *Structure*" never got written ("Interviews with Tom Kuhn," 48–49).

29. After several attempts to clarify what he meant by *paradigm*, Kuhn abandoned this term and tried to replace it, in turn, with *exemplar, disciplinary matrix,*

and *lexical structure*. I will not use *paradigm* when discussing his post-*Structure* work, although I believe that, with some clarifications, it remains an indispensable concept for Kuhn's philosophy, not fully replaceable by any of the expressions he subsequently used. I argued for this in *Kuhn's Legacy: Epistemology, Metaphilosophy, and Pragmatism* (New York: Columbia University Press, 2017), 19–20.

30. For discussion of the early responses to Kuhn, see Alexander Bird, *Thomas Kuhn* (Princeton, NJ: Princeton University Press, 2000); Wes Sharrock and Rupert Read, *Kuhn, Philosopher of Scientific Revolutions* (Cambridge: Polity Press, 2002); K. Brad Wray, *Kuhn's Evolutionary Social Epistemology* (Cambridge: Cambridge University Press, 2011); Mladenović, *Kuhn's Legacy*.

31. See, for example, Imre Lakatos, "Falsification and the Methodology of Scientific Research Programmes," in *Criticism and the Growth of Knowledge*, ed. Imre Lakatos and Alan Musgrave (Cambridge: Cambridge University Press 1970), 91–196; Israel Scheffler, *Science and Subjectivity*, 2nd. ed. (Indianapolis: Hackett, 1982); W. H. Newton-Smith, *The Rationality of Science* (Oxford: Oxford University Press 1981).

32. Kuhn, *Structure*, 111, 121.

33. This periodization is merely a useful heuristic for understanding the problems Kuhn was trying to solve on the path from *Structure* to the project of *Plurality*; it is thus drawing a blurry line, leaving significant overlaps.

34. See especially the 1969 postscript to *Structure*, 2nd ed., and "Logic of Discovery or Psychology of Research?" and "Reflections on My Critics," both in Lakatos and Musgrave, *Criticism and the Growth of Knowledge*, 1–23 and 231–78. The whole collection *The Essential Tension* is important for this period of Kuhn's thought, but chap. 13, "Objectivity, Value Judgment, and Theory Choice," is perhaps the most significant.

35. Kuhn thought of accuracy, consistency, simplicity, fruitfulness, and scope as key values, accepted and employed by all scientists of all times, but interpreted and ranked differently.

36. Kuhn, *Structure*, 1.

37. A historicist approach to philosophical problems and views situates them within a broader historical context. Historicism is thus holistic, always contextual, and interested in tracing emergence, development, and disappearance of the phenomena it studies. It is often narrative in form, and tolerant of polysemy. Critics would insist that it is also a version of epistemic, semantic, and even ontological relativism. For a detailed account of historicism in philosophy of science, see Thomas Nickles's entry "Historicist Theories of Scientific Rationality" in *The Stanford Encyclopedia of Philosophy* (Spring 2021 Edition), ed. Edward N. Zalta. Kuhn is the main proponent of a historicist theory of scientific rationality, as Nickles points out. Of course, the view that Karl Popper calls *historicism* does not belong in this category. By *historicism* Popper means the view that historical development is inevitable and goal-directed; historical actors can choose to be on the side that will ultimately win, or to oppose it, but they cannot change the predetermined, teleological historical development. Popper was a strong opponent of historicism, so understood. See his *The Poverty of Historicism*, 2nd ed. (London: Routledge, 1961).

38. The expression *retrospective ethnography* was famously introduced by the historian Keith V. Thomas, in his "History and Anthropology," *Past and Present* 24 (April 1963): 3–24. See Kuhn's third Shearman lecture for his comparison of historical and ethnographic research.

39. Kuhn influenced many post-Mertonian sociologists of science: Barry Barnes, David Bloor, Harry Collins, Bruno Latour, Trevor Pinch, Steve Shapin, and Steve Woolgar, among others.

40. See especially his "The Trouble with the Historical Philosophy of Science," reprinted as chap. 5 in *Road Since Structure*, 105–20. See also "A Discussion with Thomas S. Kuhn" in the same volume, 253–323.

41. Kuhn always acknowledged the importance of presentist historiography in the education of scientists: hermeneutic history of science, requiring as it does meticulous scholarship and painstaking reconstruction of past concepts, assumptions, and beliefs, would detract from the enormous amount of work that contemporary scientists need to do in order to become experts in their fields. The novelty in the last writings is that he came to see presentist narratives as indispensable *for all*, and not only for scientists.

42. See "Afterwords" (esp. 229 and 250) and "The Road since *Structure*" (esp. 97–99), both reprinted in *Road Since Structure*.

43. The perceived tension between historicism and naturalism is of long duration, complicated by the fact that *naturalism* is used in many ways. If we think of *supernatural* as its contrast class, naturalism characterizes almost all contemporary philosophy, but most philosophers who identify themselves as naturalists would think of their work as opposed to so-called armchair philosophy and methodologically continuous with the work done in natural sciences. Naturalism has been associated with ontological and explanatory reductionism, with scientism, and with positivism. It is thus sometimes identified as a downright anti-philosophical attitude that seeks to replace difficult philosophical questions with more straightforward scientific ones, to which answers are to be given on the basis of empirical research. Insofar as Kuhn is a naturalist, however, he belongs to the more integrative, nonreductionist version of the view that traces its lineage to John Dewey.

44. Kuhn, *Plurality*, this volume, p. 114; italics mine.

45. On this issue, later Wittgenstein was the most decisive influence on Kuhn's framing his questions as he did, but Kuhn's substantive answer in *Plurality* owes less to Wittgenstein than to the results of developmental and cognitive psychology.

46. He was especially indebted to Eleanor Rosch's research on concept acquisition, although he did not subscribe to the prototype theory of concepts that she supported. See, for example, her "Natural Categories," *Cognitive Psychology* 4, no. 3 (1973): 328–50; "Wittgenstein and Categorization Research in Cognitive Psychology," in *Meaning and the Growth of Understanding: Wittgenstein's Significance for Developmental Psychology*, ed. Michael Chapman and Roger A. Dixon (Berlin: Springer, 1987), 151–66. See also excellent discussion by Hanne Andersen, Peter Barker, and Xiang Chen in "Kuhn's Mature Philosophy of Science and Cognitive Psychology," *Philosophical Psychology* 9, no. 3

(1996): 347–63; for a more extensive treatment, see their book *The Cognitive Status of Scientific Revolutions* (Cambridge: Cambridge University Press, 2006).

47. Kuhn, *Plurality*, this volume, p. 181.

48. A historian of ideas—a historian of science, for example—would come to the same realization through a careful hermeneutic recovery of now abandoned scientific lexicons.

49. Kuhn makes this point in his notes for chap. 9 of *Plurality*.

50. Kuhn thinks that natural-kind terms behave somewhat differently when they sort objects (e.g., duck, geese) and when they sort materials (e.g., wood, gold) into kinds. See *Plurality*, chap. 5, section v. He probably should have qualified this observation: while English is among the languages that mark the difference between objects and materials in this way, it is not necessary for Kuhn's view (and it is not true) that all human languages mark this difference in the same way.

51. Kuhn used the platypus as an example in "Possible Worlds in History of Science," in *Possible Worlds in Humanities, Arts, and Sciences*, ed. Sture Allén (Berlin: Walter de Gruyter, 1989); reprinted as chap. 3 in *Road Since Structure*. "What should one have said when confronted by an egg-laying creature that suckles its young? Is it a mammal or is it not? These are the circumstances in which, as Austin put it, *'we don't know what to say*. Words literally fail us.' Such circumstances, if they endure for long, call forth a locally different lexicon, one that permits an answer but to a slightly altered question: 'Yes, the creature is a mammal' (but to be a mammal is not what it was before). The new lexicon opens new possibilities, ones that could not have been stipulated by the use of the old" (*Road Since Structure*, 72).

52. Kuhn's notes for chap. 6 of *Plurality*.

53. This was to be one of the central topics of chap. 8 of *Plurality*.

54. Kuhn's notes for chap. 6 of *Plurality*.

55. Kuhn's notes for chap. 7 of *Plurality*.

56. Kuhn's notes for chap. 6 of *Plurality*.

57. In chap. 1 of *Plurality*, while giving an overview of the whole book, Kuhn announced that these two questions would be addressed in the concluding chapter. See *Plurality*, chap. 1, this volume, p. 115.

58. Both of these statements are in Kuhn, *Structure*, 121.

59. A nominalist who believes that individuals exist independently of our minds and language is, of course, not a constructivist (or idealist) about individuals.

60. "The Road since *Structure*," chap. 4 in *Road Since Structure*, 101.

61. All of our practices—from the culturally universal, such as food making, mating, or child-rearing, to the culturally and historically specific, such as voting, participating in religious rituals, or engineering—are practices that are based on recognition of *some* kinds, be they natural or artefactual, concrete or abstract.

62. Kuhn's account thus sees strong continuities among species, and does not make categorization—not even linguistic categorization—*necessarily* tied to rationality.

63. Kuhn's notes for chap. 9 of *Plurality*.

64. Kuhn's notes for chap. 9 of *Plurality*.

65. Shearman Lectures, this volume, p. 70.

66. Shearman Lectures, this volume, p. 70.

67. See the third Shearman lecture, this volume, p. 80.

68. For the classical pragmatist view of warranted assertability, see John Dewey, *Logic: The Theory of Inquiry* (New York: Henry Holt, 1938), and his "Propositions, Warranted Assertability, and Truth," *Journal of Philosophy* 38, no. 7 (1941): 169–86. Kuhn offers a brief criticism of this view in the third Shearman lecture and in chap. 1 of *Plurality*. A more detailed treatment would probably have followed in chap. 9, which promised to offer Kuhn's theory of truth.

69. Kuhn probably had in mind Hilary Putnam's view of truth, as defended in *Reason, Truth, and History* (Cambridge: Cambridge University Press, 1981), rather than the often-misunderstood view of Charles Sanders Peirce. In saying that truth is the end of inquiry, Peirce did not intend to define *truth* as "whatever the end of inquiry delivers." In fact, Peirce did not attempt to define *truth*, or the predicates *true* and *false*, at all, although some of his formulations may suggest that he did. His main concern was to illuminate *the role* the concept of truth plays in practice, especially in scientific inquiry. In his view, a collective scientific inquiry, conducted by inquirers who start from different observations and assumptions, will actually converge on what *is true* about the domain under investigation. See, e.g., Charles S. Peirce, "How to Make Our Ideas Clear," *Popular Science Monthly*, January 1878; reprinted in *Writings of Charles S. Peirce: A Chronological Edition*, vol. 3, *1872–1878*, ed. Christian J. W. Kloesel (Bloomington: Indiana University Press, 1986), 257–76; "Truth and Falsity and Error" (1901), in *Collected Papers of Charles Sanders Peirce*, vols. 5 and 6, ed. Charles Hartshorne and Paul Weiss (Cambridge, MA: Harvard University Press, 1935), 394–98.

70. In the third Shearman lecture, section 2, p. 75, of this volume.

71. In this respect (as well as in many others), Kuhn is a pragmatist.

72. In his notes for *Plurality*, Kuhn expresses his gratitude to Ian Hacking, Jed Buchwald, and Peter Galison for making this point with the greatest clarity and persuasiveness.

Scientific Knowledge as Historical Product

a. The first draft of this paper was completed in 1981. Revised versions of it were given as lectures at Brandeis University (May 30, 1984), at the University of Minnesota (October 21, 1985), and in Tokyo (May 2, 1986). It was published in Japanese translation as Thomas S. Kuhn, "Rekishi Shosan toshite no Kagaku Chishiki" [Scientific knowledge as historical product], trans. Chikara Sasaki and Toshio Hakata, *Shisō* [Thought] 8, no. 746 (August 1986): 1–18. I thank Shinko Kagaya for her invaluable help in obtaining a copy of the Japanese publication. The text published here is from the original English version, kept in the Thomas S. Kuhn Papers, MC 240, boxes 23–24, Institute Archives and Special Collections, Massachusetts Institute of Technology, Cambridge, Massachusetts.

b. The most important critics of the image of science championed by logical empiricists, in addition to Kuhn himself, were Paul Feyerabend, R. N. Hanson, Stephen Toulmin, Michael Polanyi, Alistair Crombie, Mary Hesse, Nelson
Goodman, and, of course, W. V. O. Quine, among those writing in English. Continental European philosophy of science, especially in France, was, in contrast,
for a long time deeply intertwined with history of science, especially through
the influence of Alexandre Koyré, Gaston Bachelard, and Georges Canguilhem.
For Kuhn's sense of that important difference between the two traditions, see
his interview conducted by Aristide Baltas, Kostas Gavroglu, and Vasso Kindi,
published as part 3 in *The Road Since Structure: Philosophical Essays, 1970–1993,
with an Autobiographical Interview*, ed. James Conant and John Haugeland (Chicago: University of Chicago Press, 2000).

c. For John Locke's simple ideas of sense, see his *Essay Concerning Human
Understanding*, ed. Peter H. Nidditch (1689; Oxford: Oxford University Press,
1979); for Bertrand Russell's distinction between knowledge by acquaintance
and knowledge by description, see his *Problems of Philosophy* (London: Williams
and Norgate; New York: Henry Holt, 1912); for elementary propositions in
Ludwig Wittgenstein's early work, see his *Tractatus Logico-Philosophicus*, trans.
C. K. Ogden (London: Kegan Paul, Trench, Trubner, 1922).

d. The distinction between *context of discovery* and *context of justification*
was introduced by Hans Reichenbach in *Experience and Prediction: An Analysis of
the Foundations and the Structure of Knowledge* (Chicago: University of Chicago
Press, 1938). It quickly became widely accepted by logical empiricists.

e. The term was popularized by Rudolf Carnap, who drew it from the em-
bryologist Hans Driesch; see Carnap, *The Logical Structure of the World: Pseudo-
problems in Philosophy*, trans. Rolf A. George (Berkeley: University of California
Press, 1967), 101–3. Knowledge, according to this position, is something rationally constructed from within the stream of experience. The qualifier *methodological* was, for its part, meant to signal agnosticism about the existence of
a self or a transcendental ego. The term and its individualistic connotations
were common targets for criticism, for example by Otto Neurath; see Neurath,
"Protokollsätze," *Erkenntnis* 3, no. 1 (1932): 204–14. I am grateful to Evan Pence
for this endnote.

f. The so-called Duhem-Quine thesis is the claim that theories are underde-
termined by evidence: individual empirical hypotheses cannot be confirmed or
falsified on their own, in isolation from inevitable auxiliary hypotheses. Classic
statements of the position may be found in Pierre Duhem's *The Aim and Structure
of Physical Theory*, trans. Philip Paul Wiener (Princeton, NJ: Princeton University
Press, 1954), chap. 6; and in W. V. O. Quine's "Two Dogmas of Empiricism," *Philosophical Review* 60, no. 1 (1951): 20–43; reprinted in Quine's *From a Logical Point of
View* (Cambridge, MA: Harvard University Press, 1953), 20–46.

The Presence of Past Science (The Shearman Memorial Lectures)

a. Kuhn is likely responding here to Brent Berlin and Paul Kay's important
work on color grouping across cultures and languages. Although it may seem

that we could sort colors in a number of different ways, all of them arbitrary, Berlin and Kay argue that prototypical color terms found in different languages tend to cluster in similar ways. See Berlin and Kay, *Basic Color Terms: Their Universality and Evolution* (Berkeley: University of California Press, 1969). I am grateful to Evan Pence for this reference.

b. Kuhn is probably thinking here of the Wittgenstein-inspired prototype theory of concepts, developed by the psychologist Eleanor Rosch. See, for example, Eleanor Rosch, "Principles of Categorization," in *Cognition and Categorization*, ed. Rosch and Barbara Bloom Lloyd (Hillsdale, NJ: Lawrence Erlbaum Associates, 1978), 27–48; Eleanor Rosch and Carolyn B. Mervis, "Family Resemblances: Studies in the Internal Structure of Categories," *Cognitive Psychology* 7, no. 4 (1975): 573–605.

Abstract for *The Plurality of Worlds*

a. Summaries of the extant text of *The Plurality of Worlds* are given in the Arno Pro font. Editorial reconstructions of the main ideas for the planned but unwritten parts of the book are given in a different font, Chapparal Pro. My reconstructions are based on Kuhn's foreshadowing in the drafted chapters of what was to come later, and on the notes that he left for each projected chapter (kindly made available by the University of Chicago Press, with the permission of Kuhn's literary executors, Jehane Kuhn and Sarah Kuhn). The amount of available information for each of the unfinished chapters is uneven. Whenever possible, I tried to give the reader a sense of the general direction that Kuhn intended to take in each chapter.

b. There are passages in Kuhn's notes in which he seems to doubt whether the no-overlap principle really applies to all singletons.

c. Kuhn gave the foregoing title, "Looking Backward and Moving Forward," to chapter 7 in the table of contents of his last draft of *Plurality*. The notes that he left for chapter 7, however, give a different title: "The Many Worlds of Kinds."

d. Kuhn's notes for this chapter do not provide sufficient basis for articulating what his answer to these two questions would have been, had he lived to complete the book. What follows are merely a few strands of his thought left in his notes, which may, or may not, have been incorporated into the final version of the chapter.

The Plurality of Worlds

Acknowledgments

a. Kuhn left, among his notes for this book, a list of people to be thanked in the acknowledgments. The list does not explain specific debts, and is very probably incomplete.

Chapter 1: Scientific Knowledge as Historical Product

a. The view that truth is the end of inquiry is widely attributed to Charles Sanders Peirce. See, for example, "How to Make Our Ideas Clear," *Popular Science*

Monthly, January 1878; reprinted in *Writings of Charles S. Peirce: A Chronological Edition*, vol. 3, *1872–1878*, ed. Christian J. W. Kloesel (Bloomington: Indiana University Press, 1986), 257–276.

b. For the classical pragmatist view of warranted assertability, see John Dewey, *Logic: The Theory of Inquiry* (New York: Henry Holt, 1938), and his "Propositions, Warranted Assertability, and Truth" (*Journal of Philosophy* 38, no. 7 (1941), pp. 169–86).

c. In his notes for chapter 1, Kuhn offers a response to the question of how to preserve the necessity of noncontradiction without endorsing a correspondence theory of truth. Comparative judgments do not require a scale with truth as a limit. They require "an existential choice." An existential choice is subject to a version of the law of noncontradiction: although we can choose any candidate (a spouse, a paradigm, a language . . .), we cannot choose more than one at a time. Specifically, a single scientific community faced with two incompatible lexicons cannot accept both of them at the same time. We don't need reference to truth in order to explain such historically situated choices. Kuhn credits his wife, Jehane Kuhn, with the central insight of this response.

Chapter 3: Taxonomy and Incommensurability

a. Pluto was still considered a planet at the time of Kuhn's death. The reclassification of Pluto as dwarf planet happened on August 24, 2006, by majority vote of the International Astronomical Union.

b. Part II of the book, entitled "A World of Kinds" and consisting of chapters 4–6, remains unfinished.

c. "Member of a singleton" sounds odd. Kuhn probably should have said *instantiation* or *application* of a singleton.

d. The Italian phrase *traduttore traditore* may be rendered as "A translator is a traitor."

Chapter 4: Biological Prerequisites to Linguistic Description: Track and Situations

a. Kuhn's note in the margin: "This seems to me now clearly wrong. Must use nomic/normic distinction only within worlds." Kuhn introduced the distinction between nomic and normic in his "Afterwords," in Horwich, *World Changes*, 311–41 (Cambridge, MA: MIT Press, 1993); reprinted as chap. 11 in *Road Since Structure*.

Chapter 5: Natural Kinds: How Their Names Mean

a. Here, Kuhn made the following note to himself: "Try to cite paper by Daiwie Fu on what you get by studying other cultures." It is not clear whether he had in mind a published paper (and if so, which), or a draft.

b. In most classifications, fish are animals, but here Kuhn thinks of these categories as being mutually exclusive: for example, as if *animals* meant *mammals*.

Chapter 6: Practices, Theories, and Artefactual Kinds

a. Obviously, Kuhn could not have meant this literally, and probably would have been more specific in the final version of the text. We should read *bowls* and *spoons* here as merely examples of *possible* (and as it happens, in our culture, actual) contrast classes to *cups* and *forks*. The point is that mastery of artefactual kinds, no less than mastery of natural kinds, requires mastery of contrast classes. (This chapter is incomplete both in the sense of being unfinished and in the sense of being unrevised.)

INDEX

Accademia del Cimento, 121

Archimedes, 11, 93, 121, 127, 129, 155–57, 184

Aristotle, xvi; astronomy and, 36, 165; beliefs and, 155–57, 160; biology and, 30, 59n12, 136–37, 200, 232, 238n11, 263; *Categories*, 59n12, 162–63; conceptual change and, 135n5; developmental approach and, 155; *eidos* and, 34n8; essence and, 33–34, 139–40, 141n11, 155, 160, 163n35; on essential vs. accidental, 139n10; ethnocentricity and, 37, 42, 167; evaluation and, 156; falsity and, 49, 92, 97, 157–58, 160, 166; Galileo and, 22, 30, 32, 36, 84, 119, 136–37, 141n11, 154n12, 160, 165, 263–64; historicism and, xxiii–xxiv; incommensurability and, 52; lexical structure and, 22, 70, 77, 83–84; linguistics and, 17, 48, 53, 135n5, 154n22, 177, 194, 222; logic and, 30, 136, 160, 163n34, 164–65; matter and, 32–36, 48, 54, 63, 76–77, 92, 97, 120, 129, 135, 138–40, 155, 164–65, 175, 263–64; mechanics and, 30–31, 36, 136–37; motion and, 16–17, 30–36, 54–55, 73, 77, 83–84, 88, 92, 97, 135–41, 154–55, 158–65, 169, 194, 263–64; names and, 161; Newton and, 16–17, 30, 32, 35–36, 52, 55, 70, 73, 79, 97, 136–41, 154, 175, 200, 263–64; observation and, 136; ontology and, 32, 138; physics and, 16–17, 19, 30–42, 52, 73, 83, 91, 97, 135–41, 162–67, 200, 263–64; *Physics*, 31, 34n8, 36n9, 137, 138n7, 139n8, 154n22, 162–63, 164n36, 264; Plato and, 162n31, 164; realism and, 79; recognition and, 32; space and, 35–36, 160–66; static approach and, 84; taxonomy and, 17, 55, 63, 73, 77, 79, 176–77, 232; translation and, 31, 33n7, 36n9, 83, 135n5, 137, 139n10, 160–61, 162n31, 163n32, 163nn34–35; truth and, xliv, 49, 53, 70, 77, 79, 158, 165; vacuum and, 19, 35–36, 48–49, 53–54, 76, 79, 88, 91–92, 117, 120, 141, 160, 163–65; vocabulary and, 135n5, 153, 156, 163n32

artefactual kinds, 98; biology and, 263; falsity and, 264; linguistics and, 262n4; motion and, 263–64; names and, 263; observation and, 262, 264; as paradigm, xxxii–xxxiv; perception of differentiae and,